101 More Best Resumes

OTHER BOOKS BY JAY A. BLOCK AND MICHAEL BETRUS

101 Best Resumes

101 Best Cover Letters

101 More Best Resumes

JAY A. BLOCK, CPRW
MICHAEL BETRUS, CPRW

McGraw-Hill
New York   San Francisco   Washington, D.C.   Auckland   Bogotá
Caracas   Lisbon   London   Madrid   Mexico City   Milan
Montreal   New Delhi   San Juan   Singapore
Sydney   Tokyo   Toronto

Library of Congress Cataloging-in-Publication Data

Block, Jay A.
    101 more best resumes / Jay A. Block, Michael Betrus.
       p.   cm.
    Includes index.
    ISBN 0-07-032969-9 (pbk.)
    1. Résumés (Employment)   I. Betrus, Michael.   II. Title.
III. Title: One hundred one more best resumes.   IV. Title: One
hundred and one more best resumes.
    HF5383.B535   1999
808'.06865—dc21   99-18630
                                                    CIP

McGraw-Hill

A Division of The **McGraw·Hill** *Companies*

    3 4 5 6 7 8 9 0   MAL/MAL   9 0 4 3 2 1 0 9

ISBN 0-07-032969-9

Contents

Contributors

Acknowledgments

We would like to thank all the members of PARW who collectively have raised the bar of excellence in the area of resume writing and job coaching. Their contributions have made it possible for more people around the globe to find passion and purpose in their work.

We would also like to thank Betsy Brown, our editor at McGraw-Hill, for sponsoring the "101 ..." series, and enabling our message to reach career designers everywhere.

I also want to thank my two sons, Ian and Ryco, and my wife Dada for their love and support.

JAY A. BLOCK

I would like to thank my parents and brother for their continued support, as well as my wife Dawn for her support and putting up with my long nights working on the "101 ..." series.

MICHAEL BETRUS

Alphabetical Listing of Resumes

101 More Best Resumes

1

How to Use This Guide

Welcome to our second resume book. Our first resume book, *101 Best Resumes*, has been such a huge success that we have brought you this new installment in the series, along with *101 Best Cover Letters*. These books are very benefits intensive. Many books on the market have comparable inventories of resumes, but the resumes here, created by members of the Professional Association of Resume Writers and us, are more updated with the latest trends as well as classic layouts.

This book reviews the key structures of a successful resume, as well as offering new insights into the impact the Internet has had on resumes and tips on e-mailing your resume. We also present 101 new cutting-edge resumes, 10 new cover letters, and new tips on getting that dream position.

The largest portion of the book is dedicated to showcasing the best resumes that members of the Professional Association of Resume Writers have created for their clients. Every resume has been produced by a Certified Professional Resume Writer and was actually used by a client.

We have done enough research on this subject to know that most people buy a book like this for the sample resumes it provides, and the instruction that accompanies them may or may not be read. So, if you choose not to read the guidelines we have set forth, please consider the following tips in using the book:

■ Even if a particular sample resume is not in your area of expertise, we feel you will benefit by looking it over anyway. It may include an appealing format or approach you will like. For example, many different headlines and title styles are sampled.

- Take a good look at the boxes of hints given on the resumes. We've tried to make it easy for you to see the strategies the Certified Professional Resume Writer used in designing those resumes.

- Notice the relaxed writing style in the cover letters. Try not to write in too stiff or formal a manner.

Again, look at the many sample resumes provided by the Certified Professional Resume Writers. Whatever you do for a living, you should still look at the formats of **all** the resumes for ideas on layouts, different ways of writing, and the impact of including graphics and clip art in your resume. The resumes also exemplify a variety of ways that people have utilized the "Five P's" you will learn about in Chapter 6.

2

How Will You Find That Dream Position?

There are several primary sources of job leads:

- Networking
- Contacting companies directly
- Classified advertisements
- Executive recruiters and employment agencies
- On-line services

Other sources include trade journals, job fairs, college placement offices, and state employment offices. One of the most difficult tasks in life is securing work and planning a career. A career is important to everyone, so you must create a plan of action utilizing more than one of the career design strategies at the same time.

1. If you see a classified ad that sounds really good for you but only lists a fax number and no company name, try to figure out the company by trying similar numbers. For example, if the fax number is 555-4589, try 555-4500 or 555-4000. Get the company name and contact person so you can send a more personalized letter and resume.

2. Send your resume in a Priority Mail envelope for the serious prospects. It only costs $3 but your resume will stand out and get you noticed.

3. Check the targeted company's Web site; they may have job postings there that others without computer access haven't seen.

4. If you see a classified ad for a good prospective company but a different position, contact the company anyway. If they are new in town (or even if they're not), they may have other, non-advertised openings.

5. Always have a personalized card with you in case you meet a good networking or employment prospect.

6. Always have a quick personal briefing rehearsed in case you meet someone who could be helpful in your job search.

7. Network in non-work environments, such as a happy hour bar (a great opportunity to network) or an airport.

8. Network with your college alumni office. Many college graduates list their current employers with that office, and they may be good sources of leads, even out of state.

9. Most newspapers list all the new companies that have applied for business licenses. Check that section and contact the ones that appear appealing to you.

10. Call your attorney and accountant and ask them if they can refer you to any companies or business contacts; perhaps they have good business relationships that may be good for you to leverage.

11. Contact the Chamber of Commerce for information on new companies moving into the local area.

12. Don't give up if you've had just one rejection from a company you are targeting. You shouldn't feel you have truly contacted that company until you have contacted at least three different people there.

13. Join networking clubs and associations that will expose you to new business contacts.

14. Ask your stockbroker for tips on which companies are fast-growing and good companies to grow with.

15. Make a list of everyone you know and use them all as network sources.

16. Put an endorsement portfolio together and mail it out with targeted resumes.

17. Employ the hiring proposal strategy.

18. Post your resume on the Internet, selecting news groups and bulletin boards that will readily accept it and that match your industry and discipline.

19. Don't forget to demonstrate passion and enthusiasm when you are meeting with people, interviewing with them, and networking through them.

20. Look at your industry's trade journals. Nearly all industries and disciplines have multiple journals, and most journals have an advertising section in the back that lists potential openings with companies and re-

cruiters. This is a great resource in today's low unemployment environment.

21. Visit a job fair. There won't be managerial positions recruited for but there will be many companies present, and you may discover a hot lead. If companies are recruiting in general, you should contact them directly for a possible fit.

22. Don't overlook employment agencies. They may seem like a weak possibility, but they may uncover a hidden opportunity or serve as sources to network through.

23. Look for companies that are promoting their products using a lot of advertising. Sales are probably going well and they may be good hiring targets for you.

24. Call a prospective company and simply ask who their recruiting firm is. If they have one they'll tell you, and then you can contact that firm to get in the door.

25. Contact every recruiter in town. Befriend them and use them as networking sources if possible. Always thank them, to the point of sending them a small gift for helping you out. This will pay off in dividends in the future. Recruiters are always good contacts.

NETWORKING

Without question, the most common way people find out about and obtain new positions is through networking. Networking is *people connecting*, and when you connect with people you begin to assemble your network. Once your network is in place, you will continue to make new contacts and communicate with established members. People in your network will provide advice, information, and support in helping you to achieve your career goals and aspirations.

Networking accounts for up to 70 percent of the new opportunities uncovered. So what is networking? Many people assume they should call all the people they know, personally and professionally, and ask if they know of any companies that are hiring. A successful networker's approach is different.

A successful networker starts by listing as many names as possible on a sheet of paper. These can include family, relatives, friends, coworkers and managers (past and present), other industry contacts, and anyone else you know. The next step is to formulate a networking presentation. Keep in mind it need not address potential openings. In networking the aim is to call your contacts to ask for career or industry advice. The point is, you're now positioning yourself not as a desperate job hunter but as a *researcher*.

It is unrealistic to ask people for advice like this:

Mark, thanks for taking some time to talk with me. My company is likely to lay people off next month and I was wondering if your company had any openings or if you know of any.

This person hasn't told Mark what he or she does, has experience in, or wants to do. Mark is likely to respond with, "No, but I'll keep you in mind should I hear of anything." What do you think the odds are that Mark will contact this person again?

A better approach is to ask for personal or industry advice and work on developing the networking web:

Mark, Paul Jonathan at CNA suggested I give you a call. He and I have worked together for some time, and he mentioned you work in finance and are the controller at Allied Sensors. I work in cost accounting and feel you'd likely be able to offer some good career advice. I'd really appreciate some time. Could we get together for lunch sometime in the next week or so?

You have now asked for advice, not a job. People will be much more willing to help someone who has made them feel good about themselves or who appears to genuinely appreciate their help. This strategy can be approached in many ways. You can ask for: job search advice (including resume or cover letter advice); overall career advice (as shown above); industry advice; information about various companies, people, or industries; other people or key contacts the person may know. It is important that the person you network through likes you. When someone gives you a reference, it is a reflection on that person. They will not put themselves at personal or professional risk if they aren't confident you will be a good reflection on them. Finally, send each person you speak with a thank-you letter. That courtesy will be remembered in future contacts.

25 NETWORKING TIPS

1. Two-thirds of all jobs are secured via the networking process. Networking is a systematic approach to cultivating formal and informal contacts for the purpose of gaining information, enhancing visibility in the market, and obtaining referrals.

2. Effective networking requires self-confidence, poise, and personal conviction.

3. You must first know the companies and organizations you wish to work for. That will determine the type of network you will develop and nurture.

4. Focus on meeting the "right people." This takes planning and preparation.

5. Target close friends, family members, neighbors, social acquaintances, social and religious group members, business contacts, teachers, and community leaders.

6. Include employment professionals as an important part of your network. This includes headhunters and personnel agency executives. They have a wealth of knowledge about job and market conditions.

7. Remember, networking is a numbers game. Once you have a network of people in place, prioritize the listing so you have separated top priority contacts from lower priority ones.

8. Sometimes you may have to pay for advice and information. Paying consultants or professionals or investing in Internet services is part of the job search process today, as long as it's legal and ethical.

9. Know what you want from your contacts. If you don't know what you want, neither will your network of people. Specific questions will get specific answers.

10. Ask for advice, not for a job. You cannot contact someone asking if they know of any job openings. The answer will invariably be no, especially at higher levels. You need to ask for things like industry advice, advice on geographic areas, etc. The job insights will follow but will be almost incidental. This positioning will build value for you and make the contact person more comfortable about helping you.

11. Watch your attitude and demeanor at all times. Everyone you come in contact with is a potential member of your network. Demonstrate enthusiasm and professionalism at all times.

12. Keep a file on each member of your network and maintain good records at all times. A well-organized network filing system or database will yield superior results.

13. Get comfortable on the telephone. Good telephone communication skills are critical.

14. Travel the "information highway." Networking is more effective if you have e-mail, fax, and computer capabilities.

15. Be well prepared for your conversation, whether in person or over the phone. You should have a script in your mind of how to answer questions, what to ask, and what you're trying to accomplish.

16. Do not fear rejection. If a contact cannot help you, move on to the next contact. Do not take rejection personally—it's just part of the process.

17. Flatter the people in your network. It's been said that the only two types of people who can be flattered are men and women. Use tact, courtesy, and flattery.

18. If a person in your network cannot personally help, advise, or direct you, ask for referrals.

19. Keep in touch with the major contacts in your network on a monthly basis. Remember, out of sight, out of mind.

20. Don't abuse the networking process. Networking is a two-way street. Be honest and brief and offer your contacts something in return for their time, advice, and information. This can be as simple as a lunch or offering your professional services in return for their cooperation.

21. Show an interest in your contacts. Cavette Robert, one of the founders of the National Speakers Association, said, "People don't care how much you know, until they know how much you care." Show how much you care. It will get you anywhere.

22. Send thank-you notes following each networking contact.

23. Seek out key networking contacts in professional and trade associations.

24. Carry calling cards with you at all times to hand out to anyone and everyone you come in contact with. Include your name, address, phone number, areas of expertise, and specific skill areas.

25. Socialize and get out more than ever before. Networking requires dedication and massive amounts of energy. Consistently work on expanding your network.

CONTACTING COMPANIES DIRECTLY

Aren't there one or two companies you've always been interested in working for? Ideally you may know someone who will introduce you to key contacts there or inform you of future openings. The best way to get introduced to a targeted company is to have a current employee personally introduce you or make an introductory phone call for you. You could make the introduction and refer to the employee you know. We'll get into this later, but if you don't know anyone at a targeted company, a recruiter may be a good source of contact for you, even if it involves no job order for them.

You could send an unsolicited resume, but the likelihood of this being effective is low. Most large companies receive thousands of resumes a year, and few are acted on. Corporate recruiters Jackie Larson and Cheri Comstock, authors of *The New Rules of the Job Search Game*, don't regard mass-mailed resumes very seriously. Part of the problem is that too many resumes are written as past job descriptions and are not *customized* toward a targeted position.

Conrad Lee, a Boca Raton recruiter, believes "information is the most important thing in contacting companies directly. Don't call just one person in the company and feel that is sufficient. That person may have job insecurities or be on a performance improvement plan. You should contact five to ten people and only then can you say you contacted that company directly." New job search strategies all suggest targeting a select few smaller companies (under 750 employees, as larger companies are still downsizing) intensely rather than blanketing a thousand generically. Contacting the head of your functional specialty in that company is a good start. Is it hard? Of course. You're facing rejection, probably feeling like you're bothering busy people or begging, and maybe even feeling inferior. Would you feel inferior if you were calling hotels and ticket agencies for Super Bowl information? Of course not. What if some can't help you? You just get back on the phone until you achieve your goal. These contacts should be approached in the same way. You have a great product to sell—yourself. Position yourself as someone of value and as a product that can contribute to the target company.

The key is to position yourself for individual situations. This requires specialized letters, resumes, and strategies tailored for each situation.

One trick is to call the company you are targeting and try to get the name of the person in charge of the department you would like to work in. If you don't know, call the receptionist and ask her or him who that is, and perhaps who a vendor or two might be (such as an accounting firm or ad agency). Finally, check their Web site for the latest company news. Now you have something interesting to talk about when you reach the hiring manager.

25 TIPS FOR JOB SEARCHING WHILE STILL EMPLOYED

1. Do not let your current employer find out about your intent to look around. This means no loose resumes left on the copy machine, no mailing from the office, no signals that could jeopardize your current position.

2. Get organized and commit to the job search process. Without the immediate pressure to look for a job that comes from being unemployed, you may run the risk of being sporadic in your job search efforts. You must schedule time for the search and stick to it.

3. Don't feel guilty about looking around while employed. You owe it to yourself to make the most of your career, especially in today's environment of companies looking out for their own financial health.

4. Get a voice mail pager to enable yourself to return calls quickly, or a reliable answering machine at home.

5. Do not circulate your work number for new employment purposes.

6. Do not send your resume or any other correspondence on current employer stationery.

7. Take advantage of different time zones to make calls, if this applies to you. This enables you to make calls early in the morning or after work.

8. Use a nearby fax for correspondence if you don't have one at home (but not the one at work).

9. Do not use any resources of your current employer in your job search.

10. Commit to 10 to 12 hours per week for job searching, and schedule your activities for the week during the prior weekend.

11. Utilize executive recruiters and employment agencies. In some cases they will be able to significantly cut down on your leg work.

12. Target direct mail efforts on the weekdays after hours; though the success rate is lower than calling directly, it requires some less time on the weekdays (while you're working) than other activities.

13. Make use of lunchtime during the week to schedule phone calls and interviews.

14. Network through your family and friends.

15. Use electronic means to speed up your search, including surfing the Net for job listings and company information.

16. Try to schedule interviews and other meetings before the workday (e.g., breakfast meetings) and after 5 p.m. You'll be shocked at how many employers will try to accommodate this, and they'll appreciate your work ethics.

17. Though the hit rate may not be great, you may consider identifying a direct mail company to help you directly contact many companies. They could even direct fax for you, and the rates aren't usually too high.

18. Network during off hours and through a few professional contacts, using caution and good judgment as to who should be contacted.

19. If you are concerned about your current employer finding out about your search, leave them off your resume and note that fact in your cover letter.

20. Consider using a broadcast letter in lieu of a resume (see page 47).

21. In confidence, utilize vendors, customers, and other people associated with your current position, especially if you want to remain in your industry.

22. Contact your stockbroker for ideas on growing companies.

23. Create a business calling card with your name and personal contact information. Hand them out in sync with your 1 to 2 minute prepared pitch about yourself.

24. Do not speak critically of your current employer.

25. Read your newspaper from cover to cover to determine what companies are growing, not who's advertising job openings.

CLASSIFIED ADVERTISING

When you depend on classified advertisements to locate job openings you limit yourself to only 7 to 10 percent or fewer of all available jobs, plus you are competing with thousands of job hunters who are reading the same ads. Keep in mind that the majority of these ads are for lower-wage positions. Do not disregard the classifieds, but at the same time don't limit your options by relying too heavily on them. Answering ads is more effective at lower levels than higher. An entry-level position or administrative support position is more likely to be found using this method than a director's position. But it is easy to review advertisements. Check the local paper listings on Sunday, the paper of the largest metropolitan area near where you live, and even a few national papers like the *Wall Street Journal* (or their advertisement summary, the *National Business Employment Weekly*) or the *New York Times*.

You may gain company insight by looking at the ads that don't necessarily match your background. You may see an ad that says, "Due to our expansion in the Northeast we are looking for … ." You have just learned of an expanding company that may need you. Review papers that have good display ads like the *Los Angeles Times*, the *Chicago Tribune*, or any other major Sunday edition.

Tactically, here is an interesting suggestion. Use the thought process above and call the company. Many classified ads list a fax number, but no company name or main number. They encourage you to fax your resume, but not to call them. In most companies, a fax number is a derivative extension of the main number. So, if the fax number is NXX-5479, there is a good chance that the main number is NXX-5000 or NXX-5400. With that information you can call the company and hunt for information, write a more interesting and industry specific letter, and position yourself ahead of the people who didn't use this method.

When you write a cover letter, write about the company, not just "…I am answering your ad… ." Package the resume and cover letter in a U.S. priority mail envelope so it stands out, and you will be guaranteed to at least have your resume reviewed. When answering an ad, that is your first objective.

EXECUTIVE RECRUITERS AND EMPLOYMENT AGENCIES

Employment agencies and executive recruiters work for the hiring companies, not for you. There are thousands of employment agencies and executive recruiters nationwide. Employment agencies generally place candidates in positions with a salary range under $40,000. Executive recruiters place candidates from temporary service at the administrative or executive level to permanent, senior level management. Recruiters can be a great source of hidden jobs, even if they do not have a position suitable for you at a given time.

Recruiters and agencies have a greater chance of successfully locating a position for you if your professional discipline is of a technical or specific nature, such as accounting, engineering, or sales.

25 TIPS FOR WORKING WITH EXECUTIVE RECRUITERS

1. Find a comprehensive listing of executive recruiters, those in your geographic region and those national firms that specialize in your industry. The best source of this information is *The Guide to Executive Recruiters* by Michael Betrus (McGraw-Hill).

2. Retained search firms are paid in advance or on an ongoing basis for a search, usually for higher level positions or unique, hard-to-fill positions. Retained searches are conducted for positions such as a pulp and paper expert for Scott Paper or a senior level corporate officer.

3. Contingency fees are paid when the client company hires the candidate. Typical contingency searches are for positions such as sales representative or business manager.

4. Recruiters are paid by the client company, not by you. Don't misinterpret their motivation or loyalty.

5. Executive search firms never charge prospective candidates.

How Will You Find That Dream Position?

6. Prior to speaking to a recruiter, have a description of your background well thought out and rehearsed.

7. Make sure your background description addresses your education, number of years of experience in your industry, your primary professional discipline, and several key accomplishments.

8. When a recruiter asks you what you want to do, do not appear indecisive or say anything like, "I don't really know."

9. Contact every recruiter in your preferred geographic region, whether they specialize in your industry or discipline or not. They will know what is going on in your region.

10. Contact every recruiter in the country that specializes in your specific industry or discipline. If your area of expertise is finance, accounting, or sales, you may limit it to your region of the country first.

11. Get recruiters to like you. If they like you, they will help you. Also, in order to stay in contact with recruiters who currently have nothing that suits you, try to call with information or leads for *them* as a result of your own research. You'll be giving them something and they'll be more apt to take your calls and help you network.

12. If a recruiter cannot help you immediately, probe further for a networking contact. This is an extremely valuable approach to use with executive recruiters.

13. Have a good resume prepared to send to recruiters. Unless you have a specific reason not to, always send them a resume and cover letter rather than a broadcast letter. They're going to need it anyway, so don't slow down the process.

14. Extend recruiters the same respect you would a hiring manager with whom you are interviewing.

15. If you're targeting a large company that only works with recruiters, find out the name of the recruiting firm from the company. In many cases they will refer you.

16. Do not feel hesitant to ask recruiters for advice. This is their business and they deal with employment searches every day. If you develop a good rapport, invite them to lunch—your treat.

17. You are more attractive employed than unemployed because you appear to be a lower risk. However, in today's business climate recruiters understand that many good people are looking for work. A strong reference from your previous employer will help.

18. Many recruiters have a tougher time finding qualified candidates than finding job orders from companies. Use this to your advantage and try to present yourself as someone of value to the recruiter.

19. Call a recruiter first as opposed to sending an unsolicited letter. This way you can use the letter to reflect a prior phone conversation. Though it builds more work for them, recruiters we interviewed said it is to the candidates' advantage to call first.

20. Try to bring other business or company information to the recruiter. That will help your relationship.

21. Keep in mind that when negotiating your salary through a recruiter, she or he's playing both sides to come to terms. It's usually an easier process on a personal level to negotiate through a recruiter because of the buffer.

Just don't forget that the recruiter is trying to get you hired, not necessarily get you the highest salary.

22. Though there are many large national firms, don't overlook smaller recruitment operations. A small firm may be dedicated to one client company that could be your next home.

23. Contact every recruiter described as a generalist. A generalist will take on a variety of searches, one of which may fit your background.

24. Don't be shy about asking recruiters what you should expect while working with them. If they're a quality firm, they'll be happy to walk you through the process.

25. Stay away from any recruiting firm that tries to prevent you from contacting other search firms. No legitimate recruiter will try to work with you on an exclusive basis.

ON-LINE SERVICES

Information technology is changing the way job seekers locate employment opportunities. Job hunters are now connecting with hiring authorities electronically. Thousands of astute individuals have already tapped into these powerful new technologies (databases, electronic bulletin boards on the Internet, and other on-line employment services) to help them achieve their career goals.

There are many books dedicated in their entirety to this subject. Many independent Web pages are dedicated to such a cause: You can set up your own Web page, directing prospective employers to it. Or look at your targeted employer's page; they frequently have job postings listed. However, the best methods of contact are still personal because frequently the best and higher-level positions are not posted. But Web pages are inexpensive to access and you may identify a lead through e-mail contact with hiring managers or other job seekers.

When on the Web, do several different searches, both for job keywords and for the industry or discipline that is your specialty. Also, contact your local newspaper for additional on-line sources specific to your metropolitan area.

25 TIPS FOR USING THE INTERNET IN YOUR JOB SEARCH

1. When typing your resume out with the intent of e-mailing, make sure it is in ASCII format.

2. Use keywords heavily in the introduction of the resume, not at the end.

3. Keywords are almost always nouns related to skills, such as financial analysis, marketing, accounting, Web design.

4. When sending your resume via e-mail in ASCII format, attach (if you can) a nicely formatted version in case the ASCII version does go through and the reader would like to see your creativity and preferred layout. If you do attach a formatted version, use a common program like MS Word.

5. Don't focus on an objective in the introduction of the resume but rather on accomplishments, using keywords to describe them.

6. Don't post your resume to your own Web site unless it is a very slick page. A poorly executed Web page is more damaging than none at all.

7. Before you e-mail your resume, experiment by sending it to yourself and to a friend as a test drive.

8. Look up the Web site of the company you are targeting to get recent information about new products, etc. and look at their job posting for new information.

9. Before your interview or verbal contact, research the company's Web site.

10. Use a font size between 10 and 14 points, make it all the same for an ASCII format resume, and don't create your resume for e-mailing with lines exceeding 65 characters.

11. In case your resume will be scanned, use white paper with no borders and no creative fonts.

12. Include your e-mail address on your resume and cover letter.

13. Don't e-mail your resume from your current employer's network.

14. Don't circulate your work e-mail address for job search purposes.

15. In the "subject" of your e-mail (just below the "address to" section), put something more creative than "Resume Enclosed." Try "Resume showing 8 years in telecommunications industry," for example.

16. For additional sources of on-line job searching, do a "search" on the Web for job searching, your targeted company, and your specific discipline for additional information.

17. Be careful of your spelling on the Net. You will notice more spelling errors on e-mail exchanges than you will ever see in mailed letter exchanges.

18. Try to make sure your resume is scannable. This means it uses a simple font, has no borders, no creative lining, no boldface, no underlining, no italics and limited or no columning. Though the practice of scanning is overestimated, it should still be a consideration.

19. Purchase or check out of a library an Internet directory listing the many links to job opportunities out there. There are thousands.

20. If you are using e-mail for your cover letter, keep it brief. If the reader is reading on-screen, the tolerance for reading long passages of information is reduced dramatically.

21. Always back up what you can on a disk.

22. If you post your resume to a newsgroup, make sure that is acceptable to avoid any problems with other participants.

23. Remember that tabs and spaces are the only formatting you can use in ASCII files.

24. Make sure you check your e-mail every day. If you are communicating via the Net, people expect a prompt return.

25. Don't send multiple e-mails to insure one gets through. Try to send it with a confirmation of receipt, or keep a lookout for a notice from your ISP that the message didn't go through.

3
What Skills Do You Possess?

Have you ever known a highly successful sales professional who didn't have a firm grasp and knowledge of his or her product? Ask experienced salespeople what the secret to success is and they'll say that it's knowing the product, knowing the customer, and matching the benefits of the product to the needs of the customer. This is a powerful success formula.

The job search is a sales and marketing endeavor. There is simply no way around this: *You* are the product, *you* are the salesperson, and *you* must define your customers and promote yourself to them. So, like the highly successful salesperson, the key to your success is to know your product (you) inside and out, and match the benefits of the product to the needs of your potential customers (prospective employers). In sales, we call this selling features and benefits. You must know the features of the product, known as *marketable skills*, and determine what specific benefits result from those features that would interest a prospective employer. In other words, the only reason for someone to hire you is for the benefit you offer that person or company. If an interviewer were to ask you what your strengths are, what skills you bring to the table, or what contributions you feel you could make to the company, he or she is actually asking you to identify your features and the benefit that the company would realize by hiring you.

In order to effectively communicate the features and benefits of the product, namely you, you must first take an inventory of your skills. In the simplest of terms, there are three categories of skills:

■ Job-related skills

- Transferable skills
- Self-management skills

JOB-RELATED SKILLS

There are four categories of job-related skills: 1) working with people, 2) working with data and information, 3) working with things, and 4) working with ideas. Though most of us work with all four categories at one time or another, we tend to be attracted to one or two areas in particular. Successful teachers, customer service representatives, and salespeople must be particularly skilled at working with people. Financial controllers, meteorologists, and statistical forecasters possess outstanding skills in working with data and information. Engineers, mechanics, and computer technicians enjoy using their skills to work with things; and inventors, writers, and advertising professionals must have solid creativity and idea skills.

Which category do you tend toward? Determine which job-related skills you are strongest in and which you enjoy the most. Then write a brief paragraph stating why you feel you are skilled and qualified to work in the category you selected.

TRANSFERABLE SKILLS

Transferable skills are just that—transferable from one environment to another. If you enjoy working with people, your specific transferable skills might include leadership, training, entertainment, mentoring, mediation, persuasion, public speaking, conflict resolution, or problem-solving skills. If you enjoy working with data and information, your specific transferable skills might include research, analysis, proofreading, editing, arranging, budgeting, assessing, measuring, evaluating, surveying, or pricing. If you enjoy working with things, your specific transferable skills might include knowledge of equipment, repair, maintenance, installation, set-up, troubleshooting, or building. Finally, if you enjoy working with ideas, your specific transferable skills might include creating, developing, reengineering, restructuring, painting, writing, problem solving, planning, or brainstorming.

So take fifteen minutes, sit down with a pen and paper, and write down all the skills and abilities you possess *that have value to another company*. Transferable skills are marketable and tangible qualifications that will have value to many organizations. An accountant, human resources manager, or logistics manager at General Motors has tangible transferable skills that are of value to many companies both in and out of the automotive industry.

SELF-MANAGEMENT SKILLS

Self-management skills are skills that are personality and value oriented. Self-management skills are those that describe your attitude and work ethic. They include creativity, energy, enthusiasm, logic, resourcefulness, productive competence, persistence, adaptability, and self-confidence. One cautionary note, however: *Try not to be too general in describing your self-management skills.* When you identify a specific skill, always be prepared to explain how that skill will benefit a prospective employer. For example, if you're analytical, how does that make you better prepared for a position you have designed for yourself?

When you identify and recognize your skills, you begin to know your product. If you know your product inside and out, you will never be caught off guard in an interview. In fact, you will be able to reinforce your value by emphasizing specific accomplishments you've achieved in the past, using those specific skills.

In summary, writing a powerful resume requires that you identify your marketable skills because they represent the heart of the resume. Your ability to confidently sell yourself in an interview despite stiff competition depends on knowing your skills and communicating the benefits of those skills to the interviewer. Strategic resume preparation begins with identifying what you have to offer based on where you plan to market yourself. It is the foundation for developing a powerful resume, and will be the foundation of successful interviewing as well.

4

And a Resume Is . . .

The resume is the driving force behind career design. Ironically, it's not the resume itself that is critical; it's the energy, planning, strategy, and commitment behind the resume. For a professional athlete or actor it's the preparation that makes or breaks the performance. In career design, the effort that goes into the preparation of your resume will play a major role in the outcome of your campaign. If you invest quality time and energy in developing a comprehensive and focused resume, you'll get quality results! On the other hand, if you put your resume together without much thought or reason, simply writing down your life's story and distributing it to potential employers, chances are you'll experience less than impressive results. In fact, you'll probably end up in the unenviable position of joining the 80 percent club—those who are dissatisfied with their jobs.

The resume is the driving force of career design if it is constructed in a strategic and methodical manner. With this in mind, let's define *resume*. Webster defines it as "a statement of a job applicant's previous employment experience, education, etc." This definition is hardly adequate, so let us offer you a clear and concise definition.

A resume is a formal written communication, used for employment purposes, notifying a potential employer that you have the skills, aptitude, qualifications, and credentials to meet specific job requirements. A successful resume is a marketing tool that demonstrates to a prospective employer that you can solve his problems or meet her specific needs, and thus warrant an employment interview in anticipation of being hired.

In order to demonstrate that you can meet the needs of employers, you must have specific goals and objectives. Too many job seekers have vague, ambiguous, or uncertain career goals. They say, "I want a good paying job with a progressive organization," or "I'm open to most anything." Forget that approach! You wouldn't say to a travel agent, "I'd like to go on a vacation somewhere interesting" or, "I'm open to most anything." The age-old question applies: "If we don't know where we're going, how will we ever get there, or know when we've arrived?" There is no doubt that the quality of your career—the quality of your life—is a matter of choice and not a matter of chance *only if a choice is made*.

How does all this tie in to writing resumes? There are only two types of resumes that have proven to be effective in career design, and most people use neither type. The two types are chronological and functional resume formats. If you took every resume in circulation today and put them end to end, they would circle the earth over 26 times. That amounts to about 650,000 miles of resumes. Here is a statistic that is truly astonishing: Approximately 98 percent of all resumes being circulated today don't do justice to the candidates they describe. In other words, most of the resumes are autobiographical in nature, describing just the background and experience of a candidate. The problem with autobiographical resumes is that they simply don't work.

Hiring managers and personnel professionals don't read resumes for education or entertainment. The bottom line is this: If you can identify an employer's needs or problems and *explicitly* demonstrate that you can fill those needs or effectively solve those problems, you'll be interviewed and eventually hired. It's logical and makes good common sense. In Chapter 5 we explain which resumes work, and why.

5
Successful Styles and Formats for Resumes

There are two types of resumes that are powerful and that work:

- Targeted Resumes
- Inventory Resumes

If you know the job classification or the industry or environment in which you want to work, you are a candidate for a **targeted resume**. In essence, you can identify (target) what you want to do either by job title, by industry, or both.

If you are a generalist, open to a number of options or unable to clearly identify what you want to do but able to identify your marketable skills, you are a candidate for an **inventory resume**. An inventory resume promotes one's marketable skills to a diversified audience.

TARGETED RESUME

If you know your target audience, you must create a resume that emphasizes your skills, abilities, and qualifications that *match* the needs of your target. Position the text on your resume to match the job requirements as closely as possible. For example, if you're seeking a sales position but are not fussy about what industry you sell in, you would identify the key assets and value that you bring to the table. Five such assets might be:

1. Possessing exceptional closing skills.
2. Having an active network in place that would be especially enticing to a future employer who is looking for a candidate to ignite a sudden surge of new business.
3. Training by a reputable company so a new learning curve is literally nonexistent.
4. Having a proven and verifiable track record of specific sales accomplishments.
5. Having the ability to turn around a flat and phlegmatic territory into a flourishing one.

The problem with most resumes, according to hiring authorities, is that people simply list responsibilities. You will seldom be hired based on past responsibilities, but *you have an excellent chance of being hired based on former accomplishments*.

INVENTORY RESUME

If you cannot clearly identify your target, then your resume should highlight your accomplishments and skills in a more generic manner. What benefits will a prospective employer receive in return for employing you? What skills do you bring to the table that will enhance and contribute to his or her organization?

Let's suppose you are a branch manager of a bank who is making a career change. You might have five specific skills that you can market to any number of industries, so you would develop an inventory resume with a portfolio of inventory assets that might include the following:

1. Solid sales and marketing skills.
2. Excellent financial and budgeting skills.
3. Training and development abilities.
4. Seasoned operations management skills.
5. Strong computer aptitude.

After advertising these specific skills on the resume, the balance of the document would focus on specific achievements in these five areas.

Regardless of which resume type you choose, you must incorporate pertinent information that addresses the needs, concerns, and expectations of the prospective employer or industry. Samples of both resume types are included in this book.

COMMUNICATING CRITICAL MESSAGES

A resume must communicate *critical messages*. What are critical messages? A resume is a 30-second advertisement. Understanding that, critical messages are likened to "hot buttons," using marketing terminology. Critical messages are messages that the reader of your resume needs to read. They ignite enthusiasm and eventual action—an invitation to an interview.

Career design is an exercise in self-marketing, and it's okay to be creative and to get excited about your future. When it comes to marketing yourself, there is just one ironclad rule for resume writing:

There can be no spelling or typographical errors and the resume must be well organized and professionally presented, consistent with the industry you are pursuing.

That's it! Yes, brief is better, one to two pages unless you have a very unique situation. Today, many successful career designers are incorporating graphics in their resumes, packaging them in a vibrant, exciting, and professional manner. For the first time, career designers are getting enthusiastic and excited about their resumes. After all, if *you* can't get excited about your resume, how do you expect anyone else to get excited about it?

There are two main objectives to a resume. The obvious one is that the resume is a hook and line, luring a prospective employer to take the bait and invite you to an interview. The second objective of the resume is to get you pumped up and prepare you for the interview and the process of securing a job.

RESUME FORMATS

Chronological Format

The chronological format is considered by many employment professionals and hiring authorities to be the resume format of choice because it demonstrates continuous and upward career growth. It does this by emphasizing employment history. A chronological format lists the positions held in a progressive sequence, beginning with the most recent and working back. The one feature that distinguishes the chronological format from the others is that under each job listing, you communicate your 1) responsibilities, 2) skills needed to do the job, and, most importantly, 3) specific achievements. *The focus is on time, job continuity, growth and advancement, and accomplishments.*

Functional Format

A functional format emphasizes skills, abilities, credentials, qualifications, or accomplishments at the beginning of the document, but does not correlate these characteristics to any specific employer. Titles, dates of employment, and employment track record are deemphasized in order to highlight qualifications. *The focus is squarely on what you did, not when or where you did it.*

The challenge of the functional format is that some hiring managers don't like it. The consensus seems to be that this format is used by problem career designers: job hoppers, older workers, career changers, people with employment gaps or academic skill-level deficiencies, or those with limited experience. Some employment professionals feel that if you can't list your employment history in a chronological fashion, there must be a reason and that reason deserves close scrutiny.

Combination Format

This format offers the best of all worlds—a quick synopsis of your market value (the functional style), followed by your employment chronology (the chronological format). This powerful presentation first addresses the criteria for a hire—promoting your assets, key credentials, and qualifications, supported by specific highlights of your career that match a potential industry or employer's needs. The employment section follows with precise information pertaining to each job. *The employment section directly supports the functional section.*

The combination format is very well received by hiring authorities. The combination format actually enhances the chronological format while reducing the potential stigma attached to functional formats. This results when the information contained in the functional section is substantive, rich with relevant material that the reader wants to see, and is later supported by a strong employment section.

Curriculum Vitae (CV)

A curriculum vitae (CV) is a resume used mostly in those professions and vocations in which a mere *listing of credentials* describes the value of the candidate. A doctor, for instance, would be a perfect candidate for a CV. The CV is void of anything but a listing of credentials such as medical schools, residencies performed, internships, fellowships, hospitals worked in, public speaking engagements, and publications. In other words, *credentials do the talking*.

The Resumap

The resumap is a new format that clearly breaks with tradition. The writing of the resume is a left-brain exercise in which thoughts occur in a rational, analytical, logical, and traditional manner. By engaging the right brain in this endeavor (the creative, imaginative, and stimulating side of the brain), the resume becomes a more dynamic document. An example of the resumap is found on page 99.

Resumes Designed for the Internet

Resumes designed for use on the Internet have many similarities to those designed for paper presentation. The core components of a resume, including whether an inventory or targeted approach, which format, etc., still apply. The two major differences are the need for keywords and the required absence of creative graphics. All the other parameters hold true for electronic resumes. The specifics of the electronic resume are discussed in Chapter 7.

HOW TO SELECT THE CORRECT FORMAT

Consider using a chronological format if you have an impeccable work history and your future ties to your past. Contemplate using a combination format if you have few deficiencies in experience, education, or achievements. Consider a functional format if you are a student, returning to the workforce after an extended absence, changing careers, have worked many jobs in a short period of time, have had employment breaks, or have any other history that would make using a chronological or combination format challenging. Feel free to use a CV if your credentials speak for themselves and no further information is required of you until the interview. Use a resumap when you want to be different and make a statement.

If you need to send your resume over the Internet, you may need to adopt an electronic resume layout. You could save the standard style of resume as an attachment and send it via e-mail, but it may not be able to be read by the receiver. Even different versions of MS Word cannot read each other's documents without a conversion. If you do convert, key styles and graphics may be lost, which would do more damage than good. Use the electronic style when you're sending your resume across the Internet.

In the end, exercise common sense and design a resume that best promotes *you*. There are no rules, only results. Select the format that will afford you the best chance of success.

25 RESUME WRITING TIPS

1. Allow absolutely no spelling, grammar, punctuation, or typographical errors in your resume.

2. Know your audience before you begin to prepare the document. Then write the resume for your defined audience.

3. The resume must match your skills and abilities to a potential employer's needs.

4. A resume must address your *market value* and, in 20 seconds or less, answer the question, "Why should I hire you?"

5. Key in on accomplishments, credentials, or qualifications.

6. *Sell features and benefits*. What skills do you possess and how will they contribute to the targeted organization's goals and objectives?

7. Avoid fluff. Ambiguities and generalities represent fluff; they render a resume inept.

8. Be different, courageous, and exciting. Boring resumes lead to boring jobs.

9. Package the resume in an exciting way.

10. Be sure the resume is well organized.

11. The resume must be professionally presented, consistent with the industry you are pursuing.

12. Your resume can have a distinct personality to it. Choose your language carefully; it can make a world of difference.

13. A chronological resume format emphasizes employment in reverse chronological order. Begin with your most recent job and work back, keying in on responsibilities and specific achievements. Use this format when you have a strong employment history.

14. A functional format hones in on specific accomplishments and highlights qualifications at the beginning of the resume, but does not correlate these attributes to any specific employer. Use this format when you are changing careers, have employment gaps, or have challenges to employing the chronological format.

15. A combination format is part functional and part chronological and is a powerful presentation format. At the beginning of the resume you'll address your value, credentials, and qualifications (functional aspect), followed by supporting documentation in your employment section (chronological component).

16. A curriculum vitae is a resume format used mostly in professions and vocations in which a mere listing of credentials describes the value of a candidate. Examples include actors, singers, musicians, physicians, and possibly attorneys or CPAs.

Successful Styles and Formats for Resumes

17. The five major sections of a resume are: 1) Heading, 2) Introduction, 3) Employment Section, 4) Education Section, and 5) Miscellaneous Sections.

18. Miscellaneous sections can include Military Service, Publications, Speaking Engagements, Memberships in Associations, Awards and Recognition, Computer Skills, Patents, Languages, Licenses and Certification, or Interests.

19. Write the resume in the third person and avoid using the pronoun *I*.

20. Salary history or compensation requirements should not appear in the resume. The cover letter is the place for this if it needs to be addressed at all.

21. Always include a cover letter with your resume.

22. If you are a graduating student or have been out of the workforce for a while, you must make a special effort at displaying high emotion, potential, motivation, and energy. Stress qualitative factors and leadership roles in the community, on campus, or elsewhere. By employing a degree of creativity and innovation in your career design campaign, you are communicating to a hiring authority that you can be resourceful, innovative, and a contributing team member.

23. Employment gaps, job hopping, and educational deficiencies can be effectively handled by using the combination format (or functional format).

24. The resume should be a positive document. It must tell the truth, but not necessarily the whole truth. Don't lie, but you need not tell all, either. Keep negative thoughts and concepts out of your resume.

25. The shorter the better—your resume should be only one to two pages in most cases.

6

The Five P's of Resume Writing

Now it's time to review the five P's of an explosive resume:

- Packaging
- Positioning (of information)
- Punch, or Power Information
- Personality
- Professionalism

PACKAGING

Packaging is a vital component to sales success. Most people wouldn't think of purchasing something from a store if the packaging was slightly broken. Paper stock, graphics, desktop publishing, font variations, and imaginative presentations and ideas are part of the packaging process. Most resumes are prepared on white, ivory, or gray paper. Conforming may be a recipe for disaster, so make your resume "professionally" stand out from the crowd. You'll want to remain professional and, in some cases, on the conservative side. There are various paper styles and presentation folders that are professional but unique and that provide a competitive edge. Office supply stores and your local printer are good sources of different paper stocks.

Packaging is dramatically changed when designing an electronic resume. All the personality that a paper resume allows you to demonstrate is precluded by the ASCII standards used universally. ASCII (American Standard Code for Information Interchange) is text-only copy, so formatting such as boldface type, italics, underlining, and other graphics you will see in the resumes shown later will not show up.

In fact, those graphics will likely not only be lost but will disrupt the resume flow and make it less readable than if it were straight text copy in the first place. Tabs and spacing are the only formatting features recommended for electronic resumes. It is also recommended that you use monospaced fonts (like Courier) versus proportionately spaced ones (like Times New Roman or Arial) because ASCII text is monospaced. Finally, each line on the page should be no more than 65 characters long, because many computer screens can only display up to 70 characters per line.

Look at the resumes on pages 61, 131, and 140 for an idea of how to put together an electronic resume.

POSITIONING OF INFORMATION

Positioning means organization. Organize the data on your resume so that it's easily accessible to the reader and the reader is able to quickly grasp significant information. You need to create a section of the resume (the Introduction, discussed in Chapter 7) where the key information will be displayed. In other words, by creating a highly visible section within the resume, you manipulate the reader's eyes to hone in on information that you deem essential to getting an interview. By doing this you make the best use of the hiring authority's ten to twenty seconds of review time.

You can have the best credentials in the world, backed by a powerful personality, complemented by the strongest references, but these career-making credentials are useless if your resume is sloppy, poorly organized, and difficult to read. No matter how superior you are to your competition, a prospective employer will almost never read a poorly presented document.

PUNCH OR POWER INFO

This "P" is by far the most important. When you deliver the Punch, you deliver the information that the hiring manager wants to see. It means that you are supplying the reader with *Power Info*. Power Info is information that *matches a career designer's skills, abilities, and qualifications to a prospective employer's needs*. Quite simply, Power Info is delivering the knockout Punch, indicating to a prospective employer that you meet the criteria for hire.

The employer's task is to locate candidates whose overall credentials and background meet his or her needs. Your task is to demonstrate, in your resume and later during an interview, that you have what he or she is looking for. So the starting point of all career design resumes is projecting and anticipating hiring criteria. You need to be aware of the types of people who will be reviewing your resume. Furthermore, you must determine what kind of information he or she seeks that will provide you with a clear competitive advantage and spark enough interest in you to warrant an interview.

The challenge in writing resumes is to directly address the concerns of hiring authorities, to get into the hiring person's head. What is he or she thinking? What does he or she want? What can you show him or her that will generate a reaction? In many instances, it's specific, quantifiable achievements.

This is a good time to emphasize the importance of noting specific accomplishments on your resume. The fact that you were responsible for doing something in a past job in no way assures anyone that you were successful! If your resume is full of generalities, responsibilities, and job descriptions but lacks specific successes and achievements, how do you expect a prospective employer to differentiate your resume from the other 650,000 miles of documents? The majority of attention should be placed on your accomplishments and achievements. Responsibilities don't sell. Benefits, results, and success sell. What you were responsible for in the past has little impact on your future. *What you specifically accomplished highlights your past and determines your employment appeal.*

PERSONALITY

Hiring managers want to hire people with pleasing personalities. Your resume can have its own personality. Packaging can convey a unique personality and so can words. Through the use of sumptuous vocabulary, you can turn a rather dull sentence into a more lavish and opulent one. Substitute the word *ignited* for *increased.* Change the term *top producer* to *peak performer.* Instead of being *responsible for* something, show that you were *a catalyst for major improvements in... .*

Remember, words are power. Make use of the more than 750,000 available to you in the English language. A resume does not have to be a lackluster instrument. Lighten up and let your resume dance a bit, sing a little, and entertain the reader. By displaying a personality, you display emotion. More than any other single element, emotion sells!

With an electronic resume, you are more limited in the amount of personality your resume can exhibit. Stay strict on the simplicity of the electronic resume and try to attach a more personal resume via e-mail as a fallback. Your cover letter will also provide a small window of opportunity to demonstrate some personality.

PROFESSIONALISM

Countless hiring managers believe that how people present themselves professionally will determine how professionally they will represent the company. We purchase expensive clothing, practice good hygiene, and make sure we look our very best when going to an interview because we want to make a good, lasting, and professional first impression. The resume must do the same. Once again, you are the product and you are the salesperson. Your resume is your brochure. Would you hire yourself, based on the professionalism of your resume?

What is professionalism? Well, would you ...

- Send your resume out without a cover letter, or would you enclose a personal cover letter on matching stationery addressed specifically to a targeted individual?
- Fold your resume into thirds and stuff it in a business envelope, or would you send the resume out in an attractive flat envelope without folding it at all?
- Send the resume by regular mail, or use overnight or two-day air to make a more powerful entry into the organization of destination?

- Expect the prospective employer to call you after receiving your resume (re-active responsibility), or would you make it clear that you will telephone him or her within a week to arrange an interview (proactive responsibility)?

- E-mail your resume heavy in graphics and expect the receiver to be able to interpret it clearly, or send it as an ASCII file with the formatted version attached?

Think about these questions for just a moment. What would seem more professional to you? There is a tremendous shortage of professionalism among job seekers. Embrace professionalism and you'll discover that you'll be invited to more and more interviews. That means more opportunities to get hired.

7

Anatomy of a Career Design Resume

Regardless of the resume type you choose or the format you decide upon, there are five primary sections that make up a successful career design resume, along with numerous subsections that can also be incorporated. The five primary sections are:

1. Heading
2. Introduction
3. Employment Section
4. Education Section
5. Miscellaneous Sections

THE HEADING

The heading, also referred to as your *personal directory*, consists of your name, address (with full zip code), and phone number (with area code). If you carry a portable phone or pager or have a fax machine, you can include these phone numbers in your heading. We do not recommend that you include a work number. Many hiring managers do not look favorably upon furnishing a work number. They may conclude that if you use your present company's phone and resources to launch a job search campaign on company time, you might do the same while working for them.

There are two basic methods for setting up your heading: the traditional and creative methods. The traditional method is the centered heading. This is effective for any resume, including those that will be scanned by a computer. The creative style consists of any heading that is not centered. Look at some of the many different examples of headings in the sample resumes. For style and layout ideas, look at resumes even if they do not represent your profession.

In an electronic resume, the heading should be very simple. All in a monospace font, type out your name on the first line, your street address on the next, your city and state information on the next, then your phone number, and then your e-mail address. These five lines should be tabbed in once or twice, or left-justified. Also, do not use your current work address or current work e-mail address, and do not e-mail your resume from work.

THE INTRODUCTION

An effective, power-packed introduction consists of *two or three sections*. The introduction sets the tone of the resume and swiftly connects your area(s) of expertise with the prospective employer's needs. It must answer the initial query, "What do you want to do?" or "What value can you provide my company?"

The first section of the introduction identifies who you are and what you have to offer. It is delivered in one of the following three forms:

- Title
- Objective
- Summary

For target resumes, consider using a title or an objective. An objective should not be used if it limits your scope. For example, if you work in operations, an objective might exclude opportunities that you don't even know exist. However, if you work as a nurse or accountant, your objective may be clearly defined. When developing an inventory resume, you should incorporate a summary to kick off your introduction. The purpose of the summary is to convey the scope of your experience and background and to indicate to the reader your key strengths and areas of expertise. *The first section of the introduction must ignite initial interest and make the reader want to continue.*

In the electronic resume, the introduction is critical. It is here that you capitalize on what we call **keywords**. Keywords, usually nouns, are words that describe your achievements and career goals that are generally universal to the your industry. The prospective employer's computer can scan and pick up those keywords to match the needs of the company, and then distribute your resume internally accordingly. For example, suppose you are a marketing professional and you're sending your resume over the Net to a company for their review. Your keywords in the introduction might be: segmentation targeting; advertising; brand awareness; public relations; product marketing; pricing; packaging.

THE EMPLOYMENT SECTION

The employment section is, in the majority of cases, the most important section of your resume. (Resumes of recent college graduates are among the exceptions, when academic achievements and extracurricular activities are given more weight than employment experience.)

The employment section will have the most influence on a prospective employer in determining whether you get an interview, and ultimately, a job offer. This section highlights your professional career and emphasizes experience, qualifications, and achievements. The employment module normally begins with your most recent position and works backward. Allocate the most space to the most recent positions and less space as you go back in time. If you have a sensible and strategic reason to deviate from this guideline, and it enhances your document, go for it. Otherwise reference the following information for each employer:

1. Name of company or organization
2. City or town and state where you worked
3. Dates of employment
4. Titles or positions held

How far back do you have to go? That's entirely up to you. You do not have to go back more than 10 to 12 years unless you have a good reason to do so. For instance, if you want to get back into teaching, something you did 18 years ago but stopped to raise your children, you'll want to go back 18 years. But the rule of thumb is 10 to 12 years. For the most part, what we did 15 years ago is of less consequence to an employer than what we've done recently.

Experience is not limited to paying jobs. If applicable and advantageous, include volunteer work that enhances your candidacy. Do not include salary, reasons for leaving, or supervisor's name unless you have a very specific reason for doing so. Salary history and requirements, if requested, should be addressed in the cover letter.

How the Employment Section Should Look

When using a chronological or combination format, provide specific information for each employer you worked for and for each job you performed. Include three pieces of information for each employer or job:

1. Basic responsibilities and industry- or company-specific information
2. Special skills required to perform those responsibilities
3. Specific accomplishments

The listing of your job responsibilities should read like a condensed job description. Bring out only the highlights, not the obvious. Finally, use positive and energy-oriented words. The words you choose should reflect your energy level, motivation, charisma, educational level, and professionalism. Emotion and action sell; use action and power words.

Briefly describe any special skills you used in carrying out past responsibilities. These skills might include computers that you operated, special equipment used, and bilingual capabilities. Other examples of skills that you might have employed include problem solving, communications, organization, or technical mastery. Review your daily tasks and you'll be surprised at the skills you use every day but take entirely for granted.

The major focus of the employment or experience section should be on your specific accomplishments, achievements, and contributions. What you did in terms of day-to-day functions has little impact. *What you accomplished through those functions determines your hiring appeal.* Achievements vary from profession to profession. You need to consider:

- Revenue increases
- Reengineering successes
- Awards and recognitions
- New technology introduction
- Mergers and acquisitions
- Problems identified and resolved

- Profit improvements
- Productivity improvements
- New policies and procedures
- Start-ups and turnarounds
- Inventory reductions
- Contributions made

- Expense savings
- Systems enhancements
- Quality improvements
- Reducing employee turnover
- Adding value to the company

When using a functional format, simply list the information—company name, city or state, dates of employment, and titles—and leave it at that.

Employment (Other)

Quite possibly your employment history will go back 20 years or more. Focus the majority of your resume on the most recent 10 to 12 years, and provide a brief synopsis of the rest. You are not obligated to account for every minute of your life, so use this section to summarize activities performed many years ago if it will round out your employment background.

THE EDUCATION SECTION

The education section, like any other section, should position your credentials in the very best light without being misleading. List your highest degree first and work back. If you have attended six different colleges but have no degree, you might think that these efforts indicate that you are a lifelong learner. But it could also be interpreted as project incompletion and work against you. Think carefully, strategize, and do what's right for you.

Generally, the education section appears at the beginning of your resume if you have limited work experience. A recent high school, technical school, or college graduate will, in most cases, fall into this category. As your portfolio of experience and achievements gains momentum, the education section will drop toward the end of the resume as newly formed experiences, skills, and accomplishments begin to outweigh educational experience in the eyes of a prospective employer. Finally, if your educational credentials are seen as critical or are superior to those of competing candidates, you'll want to introduce this section early in the resume.

If you have a post–high school degree, you need not list high school credentials on the resume. A job seeker with no post–high school degree should include high school graduation on the resume. Particular details you might want to address under the heading of education include:

- Grade point average (GPA), if 3.0 or higher
- Class ranking
- Honors and awards
- Scholarships

- Intramural or varsity sports
- Clubs and special classes
- Relevant course work if directly related to your target profession (mostly for recent graduates)
- Special theses or dissertations
- Internships
- Research projects
- Extracurricular activities (tutoring, volunteer work, student activities or politics, working on school newspaper)
- Career-related jobs and activities while attending school

A few words about hiding information: There are some very imaginative methods of trying to hide weaknesses and without exception, they all fail. But that doesn't mean you should accentuate any flaws in brilliant colors for all to see. The best way to overcome a weakness is to identify a corresponding strength that will more than make up for the weakness. If you have an associate's degree in business when a bachelor's degree is required, what can you do? Pinpoint areas of experience where you've proven yourself, especially where practical experience and a proven track record stand out. Demonstrate high energy and enthusiasm and stress your commitment to give 200 percent. Also, enroll in a community school to begin earning your bachelor's degree and state this in your resume or cover letter.

You need to honestly address your weakness by demonstrating powerful strengths and assets. Even if you are successful in initially fooling a hiring authority, you can be sure that the interview will be quite uncomfortable if your resume isn't straightforward. If questions are asked that you can't answer satisfactorily, you're in for an embarrassing, defensive, and unproductive meeting. You will come across as conniving and unethical.

MISCELLANEOUS SECTIONS

Military

If you served your country and received an honorable discharge, it is fine to mention this briefly in your resume. Unless your experience in the military is directly related to the profession you are pursuing (e.g., U.S. Navy aircraft mechanic applying for a job as a mechanic with an airline), then keep it very short, one to two lines at most. The more your military background supports your future career goals, the more emphasis you should give it. Underscore key skills and achievements.

Finally, and this is very important, translate military jargon into English. Many civilian employers were never in the military and don't relate to or understand military vernacular. If you are not sure of the proper equivalent civilian terminology when translating military verbiage to business terminology, seek assistance. After going through the painstaking effort of getting a hiring or personnel manager to read your resume, you want to be absolutely certain he or she can easily understand the messages you are sending.

Interests

Interests are inserted to add a human element to the resume; after all, companies hire people, not robots. This is a section that should be kept brief, tasteful, and provocative insofar as the interviewer can use this information as an icebreaker, to set the tone of the interview. It helps to build rapport.

Obviously you will want to use an interest section when your interests match job requirements, skills, or related activities that enhance your chances of getting an interview and a job offer. A country club manager may want to include tennis, golf, and swimming as hobbies. A computer teacher may want to list reading, attending motivational workshops, and surfing the Internet as hobbies. A salesperson may want to include competitive sports because many sales managers view strong competitive skills as a valuable asset in the highly competitive sales arena.

Provide one line (no more unless you have a compelling reason to do so) of information to show the reader your diversification of interests. You might try two or three athletic interests, two or three hobbies, or two or three cultural interests. This gives the prospective employer a good profile of who you are outside of work.

Community Service, Special Projects, and Volunteer Work

Many organizations place a high degree of importance on community service. They value fund-raising efforts, volunteering time to charities, and contributing to community improvement. In many cases it is good PR and enhances a company's image in the eyes of the public. Organizations that value these activities believe in the adage that "what we get back is in direct proportion to what we give." If you feel that supporting community activities, the arts, and other such causes will enhance your overall credentials, then by all means, include them on your resume.

Professional and Board Affiliations

Memberships and active participation in professional and trade associations demonstrate to a prospective employer that you 1) are a contributing member of your profession, 2) desire to advance your own knowledge and improve your skills, and 3) are committed to the future of your vocation. Pertinent affiliations should appear in your resume.

If you sit on boards of directors, this also indicates that you are well respected in your community and that you give of your time to other organizations, be they profit or nonprofit entities. These distinguishing credits should be included in your resume.

Awards, Honors, and Recognitions

No doubt these are critical to your resume because they represent your achievements in a powerful and convincing manner. It's one thing to boast about your accomplishments—and that's good. But flaunt your accomplishments *supported by specific awards and recognitions*, and that will often be the one thing that separates you from your competition.

You can illustrate your honors and recognitions:

1. in the introduction section of your resume,
2. under professional experience, or
3. as a separate section.

Technical Expertise or Computer Skills

Incorporating a section describing your specific technical and computer skills may be an effective way to quickly introduce your skills to the reader. In a high-tech, ever-changing business environment, employers are looking for people with specific skills and, even more important, for people who have the ability to learn, adapt, and embrace new technologies.

In this section consider using short bullets so the information is easily accessible. Information and data tend to get lost and confused when lumped together in long sentences and paragraphs.

Teaching Assignments

If you have conducted, facilitated, or taught any courses, seminars, workshops, or classes, include this on your resume, whether you were paid for it or not. Teaching, training, and educating skills are in demand. They require confidence, leadership, and the ability to communicate. If you have experience in this area, consider stating it on your resume.

Licenses, Accreditation, and Certifications

You may choose to include a section exclusively for listing licenses, accreditation, and certifications. Consider using bullets as an effective way to quickly and effectively communicate your significant qualifications.

Languages

We live and work in a global economy where fluency in multiple languages is an asset in great demand. Be sure to list your language skills at the beginning of the resume if you determine that these skills are critical to being considered for the position. Otherwise clearly note them toward the end.

Personal

Personal data consists of information such as date of birth, marital status, social security number, height, weight, and sex (if your name is not gender-specific), health, number of dependents, citizenship, travel and relocation preferences, and employment availability.

Employers, by law, cannot discriminate by reason of age, race, religion, creed, sex, or color of your skin. For this reason, many job seekers leave personal information off the resume. Unless you have a specific reason to include it, it's probably a good idea to limit or eliminate most personal information. For example, if you are applying for a civil service position, a social security number might be appropriate to include on the resume. If you are applying for a position as a preschool teacher and have raised six children of your own, you may want to include this information on the resume.

Here's a good test for determining whether or not to include personal information on a resume: Ask yourself, "Will this information dramatically improve my chances for getting an interview?" If the answer is yes, include it. If the answer is no or you're not sure, omit it.

25 "WHAT DO I DO NOW THAT I HAVE MY RESUME?" TIPS

1. Develop a team of people who will be your board of directors, advisors, and mentors. The quality of the people you surround yourself with will determine the quality of your results.

2. Plan a marketing strategy. Determine how many hours a week you will work, how you'll divide your time, and how you'll measure your progress. Job searching is a business in itself, and a marketing strategy is your business plan.

3. Identify 25 (50 would be better) companies or organizations that you would like to work for.

4. Contact the companies or do some research to identify hiring authorities.

5. Define your network (see Networking Tips). Make a list of everyone you know including relatives, friends, acquaintances, family doctors, attorneys, and CPAs, the cleaning person, and the mail carrier. Virtually everyone is a possible networking contact.

6. Prioritize your list of contacts into three categories: 1) strong, approachable contacts; 2) good contacts or those who must be approached more formally; and 3) those who you'd like to contact but can't without an introduction by another party.

7. Set up a filing system or database to organize and manage your contacts.

8. Develop a script or letter for the purpose of contacting the key people in your network, asking for advice, information, and assistance. Then start contacting them.

9. Attempt to find a person or persons in your network who can make an introduction into one of the 25 or 50 companies you've noted in tip 3.

10. Spend 65 to 70 percent of your time, energy, and resources networking, because 65 to 70 percent of all jobs are secured by this method.

11. Consider contacting executive recruiters or employment agencies to assist in your job search.

12. If you are a recent college graduate, seek assistance from the campus career center.

13. Scout the classified advertisements every Sunday. Respond to ads that interest you and look at other ads as well. A company may be advertising for a position that does not fit your background but may say in the ad they are "expanding in the area," etc. You have just identified a growing company.

14. Seek out advertisements and job opportunities in specific trade journals and magazines.

15. Attend as many social and professional functions as you can. The more people you meet, the better your chances of securing a position quickly.

16. Send out resumes with customized cover letters to targeted companies or organizations. Address the cover letter to a specific person. Then follow up.

17. Target small to medium-sized companies. Most of the opportunities are coming from these organizations, not large corporations.

18. Consider contacting temporary agencies. Almost 40 percent of all temporary personnel are offered permanent positions. Today, a greater percentage of middle and upper management, as well as professionals, are working in temporary positions.

19. Use on-line services. America Online, Prodigy, and CompuServe have career services, employment databases, bulletin boards, and on-line discussion and support groups, as well as access to the Internet. This is the wave of the future.

20. If you are working on your job search from home, be sure the room you are working from is inspiring, organized, and private. This is your space and it must motivate you!

21. If your job search plan is not working, meet with members of your support team and change the plan. You must remain flexible and adaptable to change.

22. Read and observe. Read magazines and newspapers and listen to CNBC, CNN, and so on. Notice which companies or organizations are on the move and contact them.

23. Set small, attainable weekly job search goals. Keep a weekly progress report on all your activities. Try to do a little more each week than the week before.

24. Stay active. Exercise and practice good nutrition. A job search requires energy. You must remain in superior physical and mental condition.

25. Volunteer. Help those less fortunate than you. What goes around comes around.

8

Cover Letters

You must include a cover letter when sending your resume to anyone. Resumes are impersonal documents that contain information about your skills, abilities, and qualifications backed by supporting documentation. In most cases, you'll send the same resume to a host of potential employers. A resume is a rather rigid instrument, and unless you customize each document for a specific audience, the resume is for the most part inflexible.

In her 1974 publication, *How to Succeed in the Business of Finding a Job*, Phoebe Taylor provides advice on cover letters that still holds true after more than 25 years:

> *If you stop to think about it for a moment, all resumes have basic similarities. Librarians' resumes are look-alikes; accountants' resumes have much in common; and so on. To get the employer to single out the "paper you," you'll have to demonstrate some ingenuity to separate yourself from the crowd.*
>
> *The cover letter provides additional pertinent information and reemphasizes your qualifications consistent with the employer's needs. As your "personal messenger," it shows your uniqueness and your ability to express yourself on paper and gives a glimpse of your personality. Addressed to a real person, "Dear Mr. Johnson" or "Dear Ms. Winters," it becomes a personal communiqué. It proves to the reader that you made the effort and used your resourcefulness to find out his or her name and title.*

A cover letter allows you to get more personal with the reader. It is the closest you can get to building rapport without meeting in person. It is a critical component in getting an interview and, eventually, the job.

Cover letters should be brief, energetic, and interesting. A polished cover letter answers the following questions concisely and instantaneously:

1. Why are you writing to me and why should I consider your candidacy?
2. What qualifications or value do you have that I could benefit from?
3. What are you prepared to do to further sell yourself?

Cover letters work best when they are addressed to an individual by name and title. They should be written using industry-specific language and terminology. Finally, you must initiate some future action. Specifically, you want to let the reader know that you will be contacting him or her for the purpose of arranging an interview or whatever the next step will be. Be proactive! Don't expect them to call you; when possible, you should launch the next step and do so with confidence and an optimistic expectation.

ANATOMY OF A COVER LETTER

The skeletal structure for a successful cover letter is:

1. Your heading and the date
2. Person's name and title
3. Company
4. Address
5. Salutation
6. 1st paragraph: Power opening—talk about the organization, not you
7. 2nd paragraph: Purpose of this correspondence and brief background
8. 3rd paragraph: Punch the "hot button(s)"—what precisely can *you* do for *them*?
9. 4th paragraph: Closing and call to action (initiate your next move)
10. Sign-off

Consider the following quotation:

I would be lying if I told you that I read every resume that crossed my desk. But I have almost never not looked at a resume that was accompanied by a solid, well-written cover letter. The lesson here is that you must learn how to write a strong letter. A cover letter should do more than serve as wrapping paper for your resume. It should set you apart from other candidates.

This quote comes from Max Messmer, CEO of Robert Half International, Inc., one of the world's largest staffing firms. Messmer suggests that most cover letters emphasize what candidates are looking for and not enough about the contributions a candidate can make to an organization. Therefore, when you are composing your letters, avoid overusing the pronoun *I*, and focus instead on the contributions you will make to the company. Don't rehash what you deliver in the resume. Whenever possible, mention information that reflects your knowledge of the organization you are writing to or the industry as a whole. Bring current news or events into the letter that will show the reader you are up to date and current with industry trends.

Chapter 9 includes ten sample cover letters for you to review. Look at each one and notice how many are written less formally than you might expect and allow the writer more creativity than a resume might. Try not to write the cover letter in too formal a style. Many entry-level candidates tend to write very stiff "professional" letters that prevent the reader from getting to know them.

YOUR COVER LETTER AS AN E-MAIL

If you are sending an electronic or e-mailed resume, your e-mail will open with a note. This is your new cover letter. The differences in a cover letter sent online are probably more pronounced than the change in your resume. E-mailed cover letters should be considered short introductions, not nearly as long as what you might write out and put in the mail.

Your cover letter might also be used as a source of keywords for computer searches. Like your resume, it should include at least a few keywords (nouns) that will give the reader an overview of your qualifications.

The same general purposes of a cover letter apply to an e-mailed one. However, it is critical to keep it short and maintain the same formatting guidelines assumed for the electronic resume. For an example of e-mailed cover letters, please refer page 49.

THE BROADCAST LETTER

There are times when a career designer is gainfully employed, content with the job, but restless enough to want to explore alternatives. Maybe you're bored, not earning what you feel you deserve, foresee trouble ahead, or just want a career change to try something different. The challenge in this situation is that you don't want to take the chance of your current employer finding out that you're looking for other work. That could cause really big trouble. The day you send out your first resume, you risk exposure. You can never be 100 percent sure where your resume will end up. Consequently, the moment you broadcast to anyone that you are exploring employment opportunities, you run the risk of exposing this to your present employer. The broadcast letter is a means to protect you to a certain degree, though even the broadcast letter is not foolproof.

The broadcast letter can also be used by those who have had sixteen jobs in the past three years, who take time off from work on occasion, or who are returning to the workplace following an extended absence. The broadcast letter becomes half cover letter, half resume. Though you'll need a resume sometime down the road, sending a broadcast letter is a technique used to attract initial attention without providing extensive detail or exposing information you'd rather not divulge at this time. Some career designers use this letter format because they feel people are more apt to read it than a resume. For instance, secretaries who screen incoming mail may not screen out broadcast letters as quickly as they do resumes.

Broadcast letters provide an effective means for discreetly communicating your employment intentions to executive recruiters or employment agencies or for informing key people in your network of your goals and objectives. The broadcast letter, by definition, *broadcasts your strengths and abilities in more depth than a cover letter but in less detail than a resume*. There are many advantages to sending out a broadcast letter. With one, you can:

- Avoid chronology of employment.
- Provide a partial listing of former employers.
- Communicate that you are presently employed and are, therefore, uncomfortable in advertising your present employer until there is interest in you as a viable candidate.
- Speak about your strong employment record and accompanying assets without mentioning educational credentials that may be viewed by others as weak.
- Overcome a challenging past, including alcohol or substance abuse difficulties, time spent in jail, physical or emotional challenges, or other similar obstacles.

A broadcast letter can be an effective way to introduce yourself and spark interest in your candidacy. You must be prepared, however, to address any challenges in subsequent communications with employers who show an interest in you after having reviewed the broadcast letter.

OTHER COLLATERAL MATERIALS

Personal Calling Cards

It may not be practical to carry your resume everywhere you go or to every meeting or event you attend. But everywhere you go and at every meeting or event you attend, you should be networking. If you connect with an individual who might be of some assistance to your career design efforts, you must be prepared to leave a calling card. We highly recommend that you have 500 to 1,000 personal calling cards printed (they are not expensive). Make it a point to hand out 100 to 150 a week for starters! Include just basic information including your name, address, phone number, and career objective or short summary of qualifications.

Thank-You Notes

You should send thank-you notes to every person who makes even the most infinitesimal impact on your career design. Stock up on some stylish, classy notecards because even a small item like a thank-you note can make a huge difference in the outcome of your efforts.

25 TIPS FOR WRITING COVER LETTERS

1. Use customized stationery with your name, address, and phone number on top. Match your stationery to that of your resume—it shows class and professionalism.
2. Customize the cover letter. Address it to a specific individual. Be sure you have the proper spelling of the person's name, his or her title, and the company name.
3. If you don't wish to customize each letter and prefer to use a form letter, use the salutation "Dear Hiring Manager." (Do not use "Dear Sir." The hiring manager may be a woman.)
4. The cover letter is more informal than the resume and must begin to build rapport. Be enthusiastic, energetic, and motivating.
5. The cover letter must introduce you and your value to a potential employer.
6. Be sure to date the cover letter.

7. An effective cover letter should be easy to read, have larger typeface than the resume (12 point type is a good size), and be kept short—4 to 5 short paragraphs will usually do the job.

8. Keep the cover letter to one page. If you are compelled to use two pages, be sure your name appears on the second page.

9. The first paragraph should ignite interest in your candidacy and spark enthusiasm from the reader. Why is the reader reading this letter? What can you do for him or her?

10. The second paragraph must promote your value. What are your skills, abilities, qualifications, and credentials that would meet the reader's needs and job requirements?

11. The third paragraph notes specific accomplishments, achievements, and educational experience that would expressly support the second paragraph. Quantify these accomplishments if possible.

12. The fourth paragraph must generate future action. Ask for an interview or tell the reader that you will be calling in a week or so to follow up.

13. The fifth paragraph should be a short one, closing the letter and showing appreciation.

14. Demonstrate specific problem-solving skills in the letter, supported by specific examples.

15. Unless asked to do so, don't discuss salary in a cover letter.

16. If salary history or requirements are asked for, provide a modest window (low to mid thirties, for example) and mention that salary is negotiable (if it is).

17. Be sure the letter has a professional appearance.

18. Be sure there are no spelling, typographical, or grammatical errors.

19. Be sure to keep the letter short and to the point. Don't ramble on and on.

20. Do not lie or exaggerate. Everything you say in a cover letter and resume must be supported in the eventual interview.

21. Be careful not to use the pronoun *I* excessively. Tie together what the company is doing and what their needs might be. To come full circle, explain how you fit into their strategy and can close potential gaps in meeting their objectives.

22. Avoid negative and controversial subject matter. The purpose of a cover letter and resume is to put your best foot forward. Negative material (job hopping, prior termination, etc.) can be tactfully addressed in the interview.

23. If you are faxing the cover letter and resume, you need not send a fax transmittal form as long as your fax number is included in the heading along with your telephone number.

24. To close the letter, use *Sincerely, Sincerely yours, Respectfully,* or *Very truly yours.*

25. Be sure to sign the letter.

9

Ten Action-Oriented Cover Letters and Broadcast Letters

Marty Zajic

9001 Flower Mound Drive, Flower Mound, Texas, 75299 (972) 555-5497

June 29, 1999

Mr. Bryan Matthews, Vice President
Oracle Systems
777 Stemmons Freeway, Suite 1400
Irving, TX 75333

Dear Mr. Matthews:

Oracle Systems is one of the fastest-growing companies in the United States. I applaud the tremendous work you are doing in the information systems services area. Your achievements, showcased as the cover story in the latest *Computer Telephony Integration,* are impressive, and I for one would like to be a contributing member of your professional team.

I understand from the article that you are looking to become the number one managed services provider in the country, and I feel the balance of my technical background and sales abilities can help tip the scales from EDS to Oracle. I offer you:

- **6 years of experience in information systems**
- **Well-versed in telecommunications, including both voice and data networks**
- **Strong market analysis and strategic planning skills**
- **A personable, team-spirited professional with a strong network (national) in place**

If possible, I would like to visit and personally meet with you to introduce myself and my qualifications. I will take the liberty of calling you next week to arrange a meeting.

Thank you for your time and consideration. I look forward to speaking and meeting with you soon.

Sincerely,

Marty Zajic

E-MAILED COVER LETTER

Dear Ms. Goff:

My experience in business management and marketing is an
excellent fit for your new telecommunications start up in
Southern California. I have watched Cox with much envy as
you launched your PCS network in other markets, and wanted
to take this opportunity to introduce myself.

Though I did not see a market development opening on your
web site, I am confident that I am a good fit and would like
to be available for consideration should a good fit come
along.

Though I am currently employed, I am very interested in Cox,
and would like the opportunity to discuss your future market
developments.

Sincerely,

David Grannan
dgrannan@cxtraitor.com
3255 Phillips Street
San Juan Capistrano, CA 90299
(714) 555-5687

REFERRAL FROM A THIRD PARTY

Jeff Anderson
4598 St. Clair Shores
St. Clair, MI 48103
(810) 555-5873

October 14, 1998

Ms. Sharon Trean
Marcus Specialty Products
710 Jefferson Avenue
Detroit, MI 48820

Dear Ms. Trean:

A mutual acquaintance, Mr. Roger Smith, recommended that I contact you regarding a possible sales opportunity with Marcus Specialty Products. I have taken the liberty of enclosing my résumé for your review. Thank you in advance for your consideration.

I now realize that I have been missing "my calling." *I love sales, but have not been selling the products and services that I love.* I am a strong sales professional with solid technical skills, but have not been selling technical products. As Sales Manager for PDC (please refer to resume), I must have sent two dozen people to your company to purchase cellular phones (and they bought!), after they saw the slick phone I use that I purchased from you!

Now here's the irony—I get more excited promoting your phones than I have ever gotten from promoting anything I've ever sold—and I've been successful in all my sales endeavors! This is why I would like to pursue a sales position with Marcus Specialty Products.

I have over 20 years of successful sales experience. I offer you the following:

- ⇒ **A strong closer; excellent cold-canvassing and market development skills**
- ⇒ **A professional demeanor**
- ⇒ **A strong network of contacts in place**
- ⇒ **Enthusiasm and high energy**

Though my résumé is quite detailed, it cannot fully profile the manner in which I have been successful. This can only be accomplished in a face-to-face meeting where we can exchange information and examine whether there might be mutual interest. I will call you in the coming week to arrange an interview. Again, I thank you for your time and review, and look forward to meeting with you soon.

Sincerely,

Jeff Anderson

DAVID ROBINSON
4597 Wilkonson Drive
Pittsburgh, PA 20557
(358) 555-4511

March 12, 1999

Roberta Alexander, Director of Human Resources
American Technical, Inc.
101 Rochelle Drive
Pittsburgh, PA 20557

Dear Ms. Alexander:

I noticed your advertisement for Sales Manager in the June edition of the <u>Medical Messenger</u>. I am very interested in pursuing this opportunity and have enclosed my resume for your review.

YOUR REQUIREMENTS	MY QUALIFICATIONS
1. Minimum of 5 years' management experience in medical sales.	1. I have 8 years experience in medical sales.
2. Extensive training & coaching experience.	2. Received "Trainer of the Year" award, Bristol, Inc., 1994-95.
3. Proven ability to adapt sales programs changes.	3. Increased territorial market share 32% per year over past 8 years.
4. A solid professional who is respected industry-wide.	4. Member & Past President, PA Medical Sales Association.

I recently read in a local newspaper that American Technical, Inc. is streamlining its operations and is positioning itself to expand into the international arena. In addition to meeting the criteria you outlined in the above-mentioned ad, I speak four languages fluently and can be an asset in the area of international sales.

I will take the liberty of calling you early next week to discuss the possibility of arranging a face-to-face meeting to explore a number of ways I feel I can contribute to American Technical, Inc. Have a great day and I look forward to speaking with you next week. Thank you.

Sincerely,

David Robinson

Mary Beth Rouse

4799 E. Wickerville Road
Saginaw, MI 48569 (517) 555-5682

May 19, 1999

Ms. Jessica Lane
Dow Chemical Company
1111 Dow Center Drive
Midland, MI 48587

Dear Ms. Lane:

Please allow me to introduce myself. I am new to the Michigan area. I have worked in the chemical products industry for the last six years, and am interested in continuing in that industry here. I had spent the last seven years in Chicago, but a recent engagement has brought me to central Michigan.

I worked for ABC Chemical in Chicago, moved on to XYZ Chemical for 2 years, and have received promotions en route with each company. Considering Dow is such a prominent player in the industry, I feel lucky to have been moved here.

My background lies in the product development within the industry. I was on the market launch team that rolled out synthetic covering for wet weather shields. We gained a 17% share within 18 months of our launch, very strong by XYZ standards.

I will stop by your office next Tuesday between 2PM and 3PM to fill out your formal application. If you can take a few moments to see me at that time, I would be very grateful. I will call you on Monday to see if this can be arranged.

Thank you for your attention. I am excited about the possibility of joining Dow Chemical Company.

Sincerely,

Mary Beth Rouse

RELOCATION REQUEST

Scott Lane

1032 Old County Road, St. Petersburg, Florida 33702
814-555-5678 e-mail slane@ibm.net

August 10, 1998

James Hawthorne
Megamanufacturer, Inc.
8507 Mega Square West
Cincinnati, OH 45212

Dear Jim:

May I ask your advice and assistance?

As you know, for the last eight years I've been continuously challenged with new marketing assignments for Superskin and I've delivered impressive sales and profit gains for all of the brands I've managed. However, now that Tracy has begun serious gymnastics training at Cincinnati Gymnastics Academy, Wendy and I would prefer to relocate to your area to minimize the separation.

Jim, since you know my abilities and potential to contribute, would you take a moment to think about people I can contact at large manufacturing/consumer goods businesses in the Cincinnati area? I'm confident that I can bring to my next employer the same strong results I delivered for Superskin:

- Improved both sales and profits for several of the company's signature brands (Superskin skin creams, Ladyfair makeup, and BabySkin).
- As Brand Manager for the entire BabySkin brand, led the brand's first sales growth since its acquisition 10 years ago.
- Led the turnaround of Ladyfair's teen products, increasing sales 15% and profits $22 million.

In addition to any contacts you can suggest, I would greatly appreciate your insight with regard to the Cincinnati job market. To assist you in evaluating appropriate contacts and suggestions, I have enclosed a resume and offer the following important elements for my next position:

- Leadership of a large-scale marketing initiative, preferably in the consumer goods industry.
- The opportunity to have a significant, positive impact on the organization and its growth and direction.
- Focus on overall market strategy as opposed to "quick fixes" (though I certainly have the background to pull these off and would be glad to offer my insight to a company facing such a challenge).
- An organization that values its resources, especially people.

Thanks, Jim. I look forward to hearing your suggestions. I will follow up with a phone call to your office next week and I hope I can take you to lunch on my next visit to Cincinnati.

Sincerely,

Scott Lane
enclosure

BROADCAST LETTER

David Heintzelman
123 Dolphin Cove, Orlando, Florida 33980
(407) 555-1212 — CJHart@aol.com

August 18, 1998

Danilo Louega
Director of Human Resources
DSI Corporation
Post Office Box 1818
Orlando, FL 33988

Dear Mr. Louega:

A few months ago, I completed the sale of Thurner Industries, Inc., a company that, in four years of leadership, I successfully turned into a highly profitable and much desired operation. Although I have been offered a similar role for another subsidiary of Thurner Companies, I would like to explore career opportunities building technology-based organizations. In anticipation of opportunities you may have for a Senior Operations or Manufacturing Executive, I enclose my résumé for your consideration. Recent accomplishments include:

- Significant turnaround of Thurner Industries, resulting in a 2300% increase in profitability and successful sale to the industry leader.

- Intense process and quality control re-engineering effort, which led to ISO 9001 certification and a 90% improvement in procedure compliance.

- Launch of a massive facility expansion and operational streamlining initiative, which boosted sales 70%.

As my achievements demonstrate, one of my greatest strengths lies in my ability to take a new or floundering operation and nurture it quickly into profitability. Throughout my career I have successfully applied the principles of growth management, staff development, and business administration to real-life corporate issues. The cornerstones of my management philosophy are excellent communication, team spirit, training, and motivation.

Be advised that my recent compensation has averaged $200,000+, but my requirements are flexible, depending upon location, job responsibilities, and other factors.

As a follow-up to this correspondence, I will call you next week to determine if my qualifications meet your needs at this time. As I have not yet discussed my plans with Thurner, I would appreciate your discretion in this matter.

Sincerely,

David Heintzelman

Enclosure

RELOCATION

Lucille Moisan

118 Hillside Drive, Olympia, WA 00879, (800) 555-1212

August 18, 1998

Stephanie Brown, B.S.N.
Director of Nursing
Seacouver Regional Medical Center
123 Main Street, South
Seacouver, WA 00891

Dear Ms. Brown:

You and I had the opportunity to meet at a regional pediatrics conference eight months ago at the Hyatt Regency in downtown Olympia. During one of the lunch breaks, we discussed your efforts to expand the neonatal unit in Seacouver. I was excited to read of your success in securing the federal grant for which you worked so hard—congratulations!

The timing of your advertisement last week for a NICU Team Leader couldn't be better, as my family and I will be relocating to Seacouver next month. With over 10 years' experience as a NICU specialist at Olympia General Hospital, I believe my qualifications and experience align well with those you are seeking in the Team Leader position. Over the last several years, I have served as weekend shift supervisor in a 15-bed Level II NICU with full responsibility for patient care, team development, and budget management.

When I came onboard, there were numerous organizational challenges to be tackled. Most prevalent among them was the need to address declining morale and dissension among team members that had developed as a result of various personality conflicts. Within a week, I succeeded in getting team members to communicate their concerns and resolve their differences. This achievement immediately improved the staff's ability to unify and work as a cohesive team toward the common goal of providing exemplary patient care.

While this is but one of the accomplishments highlighted on my enclosed résumé, I believe it speaks to my effectiveness in a leadership role. My staff would describe me as compassionate and good-natured, but they are also quick to acknowledge my abilities in assessing a situation for what it is and implementing swift measures to resolve the issue. I am confident that I can lead your NICU team to similar success.

I will be in Seacouver on a house-hunting expedition next week and will give you a call on Wednesday to schedule an interview. I look forward to hearing more about your plans for the unit!

Sincerely,

Lucille Moisan

Enclosure

NETWORKING

MARK DALGLISH ☐ 1314 LOS COLINAS BLVD. APT. 114A ☐ IRVING, TEXAS 75033 ☐ 972.555.4819

September 12, 1998

Mr. Ron Sanders
2818 Bradford Street
San Antonio, TX 78213

Hey Ron,

Greetings from North Texas! As usual, I'm keeping my options open when it comes to job opportunities so I thought I would check on the employment market there in San Antonio. I know you have a lot of contacts with the local VM SHARE organization and there are still plenty of companies that use the old mainframes (they're calling them Enterprise Servers now).

This new perspective fits right in with my new job interests. You know I've always been interested in networking computers. I was into the Internet before it was the Internet—when it was ArpaNet and MilNet back in the late 'eighties. When we were contractors at Kelly Air Force Base, I caught the networking bug working with the Defense Data Network.

I still have a soft spot for the "care and feeding" of large mainframes and I keep my hand in with the new OS/390 operating system, and of course with VTAM and CICS. That fits right in my new bag of tricks which includes my ability to network just about every platform you know about with anything else. I'm using servers, bridges, routers, UNIX, AIX, Windows/NT, IBM ComManager, SNA, TCP/IP, LANs, WANs (Frame Relay included), satellite links and even Datalink Control Switching.

Ten, even five years ago we never realized there were so many ways to connect two, or ten, or even a hundred computers together. The thought of moving gigabytes of data between them in a few minutes was mind-boggling. Now I work in that environment every day. Who'da thought?

So—how about checking with our old buddies over at USAA and Hart-Hanks to see if they can use an old dog who's learned a bunch of new tricks. I'm enclosing a few copies of my current resume to pass around. Call me in the evening at 972.555.4819 or Email me at DArthur@aol.com if you have any hot news.

Thanks! All my best to Nikki.

Your friend,

Mark

Enclosure

REFERENCE LETTER

Sun N
Internati
Breakwat
Corpus
Texas

Surf
onal ǁ 413
er Blvd. ǁ
Christi,
73633

Ms. Lori Harding,
The University Club
409 Inwood Road
Dallas, Texas 75203

Sales Director

Dear Ms. Harding:

This letter refers to Ms. Nancy Newman, who was a valued member of the Sun N Surf International management staff for six years. She recently relocated to Dallas and called me a few days ago to express an interest in your organization. I recommend Ms. Newman unreservedly to you for the position of Pro Shop Manager.

During her tenure with Sun N Surf, Ms. Newman managed our beach club Pro Shop in a fashion I can best describe as exemplary. She attended five markets a year to continually upgrade and expand the shop's inventory of men's and women's active wear and beach apparel. Her initial efforts sparked the interest of the membership and ignited sales and new business.

As manager of a shop that attracts affluent members of the corporate world, she learned that money goes where it is well treated. This reality was incorporated into her hiring decisions (she directed a staff of five), staff training and supervision, special events, promotions and personal customer assistance. It is also reflected in the growth of our client base since she came on board.

Ms. Newman never exceeded a budget, missed a deadline or failed to meet or exceed the company's sales goals. To stay within budget parameters, she would sometimes manage two functions to eliminate excess payroll hours for the staff. She brings flair, resourcefulness and sound business sense to organizations like ours. I feel confident that she will bring you the results you are looking for.

Thank you for your time and consideration. If I can be of further assistance, feel free to write me at Sun N Surf International in Corpus Christi, or call me directly at 210-555-1274.

Sincerely,

Jason Wells
Vice President, Sales

10

101 Career Design Resumes That Will Get You Hired!

1—ACCOUNTANT/STUDENT

RUSTY WILLIAM CATES
78 Addington Place, #12, Calcu, Missouri 54321
(564) 555-3456

PROFILE	• **Academic training and interest in Accounting. Willing to travel or relocate.** **Well-versed in, and acquired hands-on experience (through college projects of:**

- financial statements & analysis
- asset, liabilities, & stockholders' equity
- cost & managerial accounting
- accounting information systems
- 10 K report & financial investors' analysis

- revenue & expenses
- pensions & leases
- tax accounting
- auditing
- consolidations

• **Computers:** IBM / Macintosh platforms - DOS - MS Word - WordPerfect - Lotus -Excel - dBase - FoxPro - Power Point - most online services - Internet.

EDUCATION	• **Bachelor of Science - Accounting,** 1995 University of Missouri, Currency, Missouri GPA: 3.6 / 4.0 ACADEMIC FOCUS: FINANCIAL & COST ACCOUNTING & AUDITING ACTIVITIES

- **Elected to leadership role and served on Finance and Social Committees** of Delta Omega Pi fraternity.
- **Volunteered time and resources to the Boy's and Girl's Club of Lancaster** on a variety of projects.
- **Participated in intramural college sports:** basketball, football, softball.

• **Currently enrolled at Calcu College of Finance**
- Twelve-hour pre-CPA Examination coursework.

HIGHLIGHTS OF EXPERIENCE

WALLACE MANUFACTURING
- **Acquired first-hand knowledge of how a business runs,** both from an operational aspect and staffing requirements.
- **Learned the importance of completing work projects on time** - accepting personal responsibility for work quality - and contributing as a productive team member to accomplish an assignment.

SPIRO'S DEPARTMENT STORE
- **Sold men's clothing and accessories** in high-volume retail environment. Determined customer needs and suggested items to complete wardrobe (and increase sales total).
- **Ensured complete customer satisfaction by providing attentive, personalized service** which resulted in building long-term customer relationships and repeat business.

XEROX CORPORATION
- **Supplied manual labor** to several company cafeterias and the plastic packaging and shipping department.
- **Maintained a positive attitude while performing routine or menial functions.**

CITY OF CALCU
- **Planned, coordinated, and administered beginning and intermediate swimming classes for children.** Oversaw all operational and safety aspects of pool facility.
- **Obtained professional certifications** to perform duties as a Lifeguard and Swimming Instructor and motivated students to overcome their fear of water.

WORK HISTORY (during college)

- WALLACE MANUFACTURING, Calcu, Missouri
  Machine Operator: Plastic Molding (9/94-10/95)
- SPIRO'S DEPARTMENT STORE, Mile High, Colorado
  Sales Associate (10/93-10/94)
- XEROX CORPORATION, Johnson, Missouri
  Food Services & Tenite Plastic Departments (summers 1992, 1993)
- CITY OF CALCU, Calcu, Missouri
  Lifeguard & Swimming Instructor (summers 1987-1991)

60

2—ACCOUNTING MANAGER/E-MAIL RESUME

Robert Atkinson
4548 Sierra Lane
Jupiter, FL 33445
(561) 555-5934
robert.atkinson@aol.com

ACCOUNTING MANAGER

PROFILE:
Passed CPA exam in State of Florida
Well versed in all relevant computer software programs,
including Peachtree, DacEasy, Lotus 1-2-3, AccountPro, Excel and
most other Windows based applications

Experience in cost accounting in manufacturing environments and
public auditing

Develop plan, conduct audits and variance analyses, process
payroll and payroll tax reports and filings, and maintain/update
accurate inventories.

KEY ACHIEVEMENTS:

- Instrumental in the negotiation and acquisition of $3 million
home care and retail pharmacy stores; negotiated a $14 million
contract for pharmaceuticals resulting in a savings of 1.5-3% on
cost of goods for each retail store.

- Saved $87k in annual corporate management salaries through
comprehensive management of the financial programs and credit
administration of the group.

- Managed accounting department consisting of controller,
billing auditor, and accounting staff.

EMPLOYMENT:

Wellington Regional Medical Center, West Palm Beach, FL
1994 - Present
Accounting Manager
- Managed accounting staff reporting to CFO

- Design, implement, and manage all centralized accounting,
management information systems, and internal control policies
and procedures

- Manage accounting department consisting of A/P, A/R, cost
management and general ledger maintenance

- Prepare all Federal and State tax requirements including
corporate, partnership, payroll, and property tax returns.

Diversified Centers, Inc., Palm Beach, FL
1988 - 1994
Staff Accountant

- Staff accountant for real estate development firm

- Coordinate all financing and external reporting with financial
institutions for the group.

- Full general ledger management responsibility

- Managed accounts payable for two years

- Managed accounts receivable for 18 months

- Prepare and administer operating and cash budgets for each
retail profit center.

EDUCATION: Florida Atlantic University, Boca Raton, FL
                  Bachelor of Arts: Accounting, 1987

> *Simple layout and font designed for E-mail and transport via the Internet.*
>
> *Simple font that is monospaced.*
>
> *No elaborate formatting, bold, underlined, or italicized fonts.*

3—ADMINISTRATIVE ASSISTANT

ALICIA PARAMO
9843 South Parkland Drive
Dallas, Texas 76545
(214) 555-9875

EXECUTIVE SECRETARY / ADMINISTRATIVE MANAGER

Ten years' experience planning and directing executive-level administrative affairs and support to Chairmen, Boards of Directors and Executive Management. Combines strong planning, organizational and communication skills with the ability to independently plan and direct high-level business affairs. Trusted advisor, liaison and assistant. Proficient with leading PC applications including word processing and presentation programs. Qualifications include:

Executive & Board Relations
Executive Office Management
Staff Training & Development
Confidential Correspondence & Data
Special Events & Project Management
Crisis Communications

> *Title helps the reader quickly grasp the position of the candidate.*
> *The listed Qualifications are also relevant and good to point out.*

PROFESSIONAL EXPERIENCE

Executive Secretary -- GLOBAL TECHNOLOGY, INC. (GTI) -- 1991 to Present

Recruited to administration/office management position and promoted to Executive Secretary in 1992 as the personal assistant to the Chairman of the Board and CEO of this diversified research venture. Executive Liaison between Chairman/CEO and Executive Management Committee, Business Departments and employees to plan, schedule and facilitate industry and inter-company business functions. Manage confidential correspondence, meetings, travel and schedule for the Chairman/CEO. Coordinate quarterly shareholders meetings, manage liaison affairs and facilitate print production of shareholder communications. Led team of 40 and managed $1.5 million budget.

- Recruited 50+ personnel responsible for all office management functions for the Chairman/CEO, Senior Vice President and CFO and CTO.
- Played key role in the successful completion of a multi-million dollar construction project to support the company's rapid growth and expansion. Negotiated contracts for the purchase of equipment and supplies.
- Exceeded all corporate standards for productivity, efficiency and operating management.

Executive Secretary / Stenographer -- UNITED STATES AIR FORCE -- 1982 to 1990

Fast-track promotion through several increasingly responsible administrative management positions throughout the U.S. and abroad. Advanced based on consistent success in effectively managing high-profile, sensitive military affairs for commanding generals and other top military officers. Received numerous commendations and awards for outstanding performance. Significant projects and achievements:

- Managed confidential correspondence, schedule and meetings for Judge Advocate. Independently researched, and responded to requests for legal documentation for a team of 10 attorneys managing 100+ cases per week.
- Planned and directed security, logistics and administrative affairs for visiting foreign dignitaries, congressional leaders and military officials.
- Selected from a competitive group of candidates for exclusive joint military Leadership Development Course.

EDUCATION & PROFESSIONAL TRAINING:

- Graduate of 200+ hour intensive U.S. Air Force executive administrative and secretarial training program.
- Attended seminars while continuing with higher education sponsored by universities and professional associations.

4—ADMINISTRATIVE ASSISTANT

Sue Ann Szumiak

38 Tamiami Trail, #22
Sarasota, FL 34242
(666) 555-1212

Qualifications Profile

- Highly motivated **Administrative Professional** with exceptional organizational skills and very methodical approach to responsibilities; exemplary abilities in managing multiple demands simultaneously.
- Highly flexible and adaptable to changing organizational needs.
- Effective problem-assessment and problem-solving skills; strong communications and editorial abilities.
- Outstanding track record of performance includes numerous commendations and achievement proclamations as well as cash awards and letters of appreciation.
- Key skill areas include word processing, expert typing and stenography (Gregg) abilities as well as extensive experience with teletype/fax and international travel reservations.

Professional Experience

1992–Present **DRUG ENFORCEMENT ADMINISTRATION** • Tampa, FL
Office Assistant (1994–Present; promotion)
- Provide a high level of administrative support including wide range of specific responsibilities for agency tasked with fighting drug war; Tampa office is largest of three offices statewide.
- Directly support Resident-Agent-in-Charge (RAC), handling preparation of all office documents.
- Author own correspondence as well as that for RAC.
- Support staff of approximately 25 undercover agents comprising drug enforcement task forces.
- Coordinate and oversee extensive travel arrangements for task force participants, including international undercover travel.
- Facilitate training of administrative staff.
- Compile complex monthly and quarterly reports; maintain professional records.
- Utilize encrypted communications system.
- Hold top secret clearance; atmosphere requires highest level of confidentiality.
- Assume responsibilities of RAC in his absence.

Group Secretary (1992–94)
- Provided key administrative and secretarial support to staff comprising 8–10 agents.
- In addition, provided timely overflow support to other personnel in absence of their administrative support staff on an as-needed basis.
- Professionally prepared investigative reports, circulating to appropriate staff.
- Ensured accurate preparation and distribution of numerous forms.
- Maintained timely personnel records as well as time and attendance reports.
- Utilized teletype and wide range of office equipment.

Education **TAMPA BAY COMMUNITY COLLEGE** • Tampa, FL
Associate's in Science Degree, Business Administration (1992)

Continuing professional development includes successful completion of a wide range of administrative and office management courses through the Drug Enforcement Administration.

Civic
- Religious Education Teacher
- Choir Member
- Blood Donor

Crisp, clean format with strong opening Qualifications Profile.

Qualifications Profile outlines key skills and abilities that will benefit a prospective employer.

Bulleted lists under each job effectively highlight skills and accomplishments.

5—ADVERTISING AND PRODUCTION

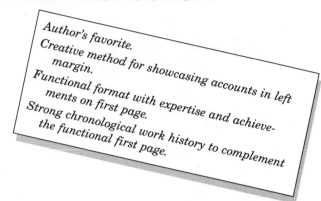

Author's favorite.
Creative method for showcasing accounts in left margin.
Functional format with expertise and achievements on first page.
Strong chronological work history to complement the functional first page.

David R. Fuery

454 East 79th Street
New York, NY 10021
Phone/Fax
(212) 555-6226

Career Profile

Fashion and entertainment industry Advertising Production Specialist. Extensive credentials in key account sales, marketing, and management.

- Produce, cast, negotiate, and budget photo shoots for multimillion dollar domestic and international accounts.
- Extensive international exposure and contacts.
- World traveled, Paris educated, working knowledge of Italian.

Expertise and Achievements

Representative Accounts
L'Oréal
Faberge
Lord and Taylor
Saks
Spiegel
Neiman Marcus
Victoria's Secret
American Express
Condé Nast
Otto Versand

Production

- Coordinate and execute worldwide photo shoots.
- Direct twenty to forty international production teams.
- Engineer and maintain high-profile fashion photographers.
- Cast, book, and schedule leading models and performers.
- Supervise departmental staff of fifteen.

Sales and Marketing

- Open and retain million dollar accounts.
- Research domestic and international industry trends, develop new contacts.
- Develop sophisticated proposals; submit bids.
- Meet rigorous deadlines, come in on budget.

Representative Agencies

Elite
Ford
Wilhelmina

Account Management

- Negotiate advertising contracts; manage million dollar accounts.
- Produce million dollar plus budgets.
- Oversee billing and talent payments.
- Business agent for clients of representation firm.

Public Relations/Representation

- Represent photographers, musicians, personalities, and screenwriters.
- Adept at subtle, successful management of creative egos.
- Liaison for international clients.
- Maintain extensive worldwide network of business contacts.

David R. Fuery

Computer Skills

- Word Perfect
- Lotus123
- Harvard Graphics
- Quicken Accounting
- Accpac General Accounting
- Boss

Employment

Senior Photographers Agent   **1993 to present**
Superior Representation L.T.D.
New York, NY

- Produce worldwide photo shoots. Scout new clients, negotiate all contracts, and maintain million dollar accounts. Maintain wide network of international business and creative contacts.

Production Manager   **1991 to 1993**
Dynamic Productions
New York, Miami, Paris

- Directed production of fifty catalog/advertising photo shoots worldwide, utilizing twenty to forty photo teams. Supervised staff of ten. Coordinated location, scheduling, budgets, and casting.

Sole Proprietor and President   **1990 to 1991**
David R. Fuery Talent, Inc.
New York, NY

- Independently engineered and established private company. Represented artistic professionals as business agent and manager.

Director   **1986 to 1990**
Fashion Model Management
New York, NY

- Booked top models for print and show. Negotiated advertising contracts, scouted worldwide, and established new client base. Supervised staff of fifteen.

Education

M.A. in Fine Arts in progress   **Pratt Institute**
New York, NY

B.A. in Marketing, 1985   **American College Paris**
Paris, France

6—APPRAISER

Robert B. Lee, MAI
115 North Union Boulevard, Colorado Springs, CO 80909 • 719/555-9050

Real Estate Professional - Residential and Commercial Appraiser

Member, American Institute of Appraisers
State of Colorado Certifications and Licenses:
Certified General Appraiser - approved to appraise any property type or any loan amount
Certified Residential Appraiser - approved for 1- to 4-unit residential properties regardless of loan amount
Licensed Real Estate Broker

SUMMARY OF QUALIFICATIONS

» Maintain all applicable licenses, certifications and continuing professional education credits
» Experienced in the appraisal of all types of properties from residential through commercial
» Unique experience in the appraisal of farm and ranch properties
» Approved to appraise for the Veterans Administration and Federal Housing Administration

EDUCATION, CERTIFICATIONS AND PUBLICATIONS

MBA, Finance, University of Colorado
Bachelor of Science, Business Administration, University of Southern Colorado
"Farm and Ranch Appraising", Appraising, Dec '96, Vol. 26, Issue 12
"Well & Septic Tank Issues", Appraising, Mar '94, Vol. 24, Issue 3

PROFESSIONAL EXPERIENCE

Self Employed Commercial and Residential Appraiser   **1994 to Present**
Colorado Springs, Colorado (and surrounding urban, rural and mountain communities)

Appraise existing and proposed residential and commercial projects for a large number of financial institutions as well as the Veterans Administration and Federal Housing Administration. Have specialized in residential development feasibility and consulting on problem real estate (both acquired through foreclosure and prior to development). Serve as Vice President of the local chapter of Real Estate Appraisers Association and as a member of the Board of Directors for the Pikes Peak Region Board of Realtors™. Directly supervise two administrative support staff. Extensive experience in appraising farm and ranch properties ranging from small acreage to large and complex cattle feeding and grazing facilities.

Chief Appraiser, First National Bank   **1987 to 1993**
Colorado Springs, Colorado

Managed all aspects of the Appraisal Department. Recruited, hired, supervised, scheduled, trained and evaluated a staff of 8 appraisers and 3 clerical. Developed and maintained all policy and procedure manuals for the department. Served as a member of the Senior Loan Committee and met regularly with members of the Board of Directors. Significant interaction with bank regulators.

Significant accomplishments:
Reduced departmental turnover from 30% to zero during tenure as Manager.
Developed computer assisted programs to track status of appraisals in process.
Maintained an "Excellent" departmental rating from Federal bank examiners.

REFERENCES, CREDENTIALS AND FURTHER INFORMATION UPON REQUEST

7—ARCHITECT

JENNIFER JOHNSON

Architect

AREAS OF PROFICIENCY

Experienced in all phases of design: from program definition through working drawings; expertise in construction estimating, cost analysis, and feasibility studies.

EDUCATION

Master of Architecture • University of California (Berkeley)	1986
Master of Building Construction • University of California (Berkeley)	1984
Bachelor of Design • University of California (Berkeley)	1981

ACADEMIC ACHIEVEMENTS

Dean's Honor List	1982–86
Design Finalist — Elderly Housing	1986
(Exhibited at National AIA Convention for Health)	
Selected Works for University Accreditation	1986
Design Finalist — Research Laboratory	1985
(Exhibited at Florissant, Colorado)	
Gargoyle Honor Society	1979–81

PROFESSIONAL EXPERIENCE

David Thomas John Rodriquez & Associates
1990–Present
Charlotte, NC   1987–89

Architect/Designer handling programming, schematic design, and design development through the use of models and drawings.

DeBoeur & Johnson, Architecture and Gardens   1989–90
Richmond, VA

Job Captain on various residential scale projects; Office Manager overseeing manpower allocation, supervision, and budgeting.

Stephen Jones & Associates, Inc.   1986
Miami Beach, FL

Job Captain on various projects of an institutional nature.

University of Florida • Gainesville, FL   1985–86
Teaching Assistantship

Donald Rosen & Associates, Inc. • Gainesville, FL   1985
Job Captain — Nursing Home

Dana Langer, Jr., Architect • Gainesville, FL   1984–85
Design Draftsman — FHA Housing

The Edward C. McCarthy Company • Miami, FL   1982–84
Senior Design Draftsman — Surveys

8—ATTORNEY

ROBIN CLATERBOUGH
567 Alamo Avenue, San Antonio, Texas 78228
Residence: 555 555-1234 • Office: 555 555-2345

ATTORNEY AT LAW

EXPERIENCE

LAW OFFICE OF GEOFFREY CISNEROS, San Antonio, Texas
Attorney-at-Law, 1993 to present
Solo practitioner with a diversified caseload of real estate transactions and litigation; business transactions and litigation; corporate and partnership matters; estate planning; probate, bankruptcy; mortgage foreclosures.

GEOFFREY CISNEROS, A Professional Corporation, San Antonio, Texas
President/Owner, 1991 to 1993
Real estate transactions and litigation; business transactions and litigation; corporate and partnership matters; estate planning; bankruptcy; mortgage foreclosures; homeowner associations.

GLADYS P. KNIGHT & ASSOCIATES, San Antonio, Texas
Associate Attorney, 1988 to 1991
Real estate transactions and litigation; business transactions and litigation; corporate and partnership matters; estate planning; bankruptcy; equine law; homeowner associations.

BLANCA, BLANCO, BLACK & WHITE, San Antonio, Texas
Associate Attorney, 1985 to 1988
Real estate transactions and litigation; business transactions and litigation; corporate and partnership matters; estate planning; bankruptcy; equine law; homeowner associations.

BLOCK & BLOOM, San Antonio, Texas
Associate Attorney, 1981 to 1985
Real estate transactions and litigation; business transactions and litigation; corporate and partnership matters; estate planning; bankruptcy; personal injury.

> *Resume effectively highlights work experience as the key criteria for future work.*
>
> *Strong Admissions and Affiliations to complement the work experience.*
>
> *One-page format that successfully blends the four sections in an organized manner.*

EDUCATION

TEXAS WESTERN SCHOOL OF LAW, San Antonio, Texas
Juris Doctor, 1978
• Phi Delta Phi International Law Fraternity

UNIVERSITY OF TEXAS, Austin, Texas
Bachelor of Arts in Political Science

PAST AND PRESENT AFFILIATIONS

Member, State Bar of Texas
Member, American Bar Association
Member, Bexar County Bar Association
Member, San Antonio County Bar Association

ADMISSIONS

Texas State Courts
Bexar County Courts

Federal Courts:
Southern District of Texas
Central District of Texas
District of Texas

Judge Pro Tempore:
Alamo Municipal Court

Arbitrator:
San Antonio Municipal Court

Appropriate personal and professional references are available.

9—AUTOMOTIVE TECHNICIAN

William Joel
111 52nd Street • Troy, Illinois 62294 •

Listing the ASE Certifications near the top helps the reader quickly see critical skills acquired. For any discipline, listing accomplishments is an important focus.

Vehicle Maintenance Technician

Extensive hands-on experience performing and supervising minor, major, and mobile maintenance for commercial/military vehicles. Outstanding troubleshooting ability and a proven track record for high levels of customer satisfaction, quality assurance, cost savings, productivity, and overall vehicle readiness. A skilled training instructor with expertise in parts management and the interpretation of vehicle system schematics.

ASE Certifications

Engine Repair   Suspension and Steering
Brakes   Electrical/Electronic Systems
Heating/Air Conditioning   Engine Performance

Professional Experience

UNITED STATES AIR FORCE, 1980 - 1998

Working in a dual role as a Mechanic and Service Manager, supervised and performed vehicle, body, and equipment maintenance to include diagnostics, repair, and rebuild of components. Provided timely customer service and ensured that repaired vehicles complied with maintenance and safety policies.

Scope of accomplishments:

- For fleets totaling as many as 450 vehicles, conducted comprehensive diagnostic and quality assurance inspections to ensure safe, serviceable, and reliable vehicles were returned to customers with 0% repeat maintenance. Established and maintained an annual privately-owned vehicle safety program, identifying numerous mechanical and safety defects to owners.
- Developed and conducted in-shop training program, increasing productivity and vehicle in-commission rates. Consistently maintained vehicle in-commission rates of about 95% and productivity rates as high as 20% above military standards.
- Established an overhaul program for vehicle transmissions, starters, and alternators, resulting in an annual savings of $4,000. Established a tire recap program saving more than $30,000.
- Raised vehicle turnaround time to new levels of efficiency.
- Identified serious frame cracks around steering gear box mount area on Chevrolet Blazers, initiating Air Force-wide inspections that revealed more extensive problems.

Education/Specialized Training

Vehicle Maintenance Management
Honor Graduate, USAF Automotive Training Course
Distinguished Graduate, USAF M.A.N. Vehicle Maintenance Course
HMMWV Organizational and General Support Maintenance Training Program

10—BANKING

JESSICA B. DALTON
2485 Adams Street
Cincinnati, OH 43002
(333) 555-1212

PROFESSIONAL SUMMARY

- **Senior Banking Professional** with a background reflecting consistent career progression and achievement of objectives. Demonstrated expertise in effective relationship development and management complemented by strong interpersonal skills. Breadth of general management experience and financial acumen attests to ability to fully utilize resources and effect results that are profitable to the bank and beneficial to the customer. Work with companies with sales of up to $10MM and loan requirements between $50K and $1MM.

- **Attributes** — Possess strong analytical and assessment skills. Keen listening and negotiating abilities facilitate understanding of all facets in decision-making process; this results in reputation for writing solid packages and good business for all parties. Proven ability to cultivate new accounts, establish strong business relationships, and immediately contribute to operations.

PROFESSIONAL EXPERIENCE

1993–96 UNITED BANK & TRUST • Cincinnati, OH
Vice President, Business Banking Group *(1995–Present, Promotion)*
Create and implement strategic business development plans overseeing operations in 7 branches supporting 2 key urban markets for commercial bank with $81 billion in assets. Manage and effectively sell full breadth of financial products and services within accounts.

- Significantly expanded bank's community development efforts through increased involvement with local, state, and federal programs (SBA); this resulted in new inroads to more complex, nontraditional business opportunities which yielded profitability as well as greater market penetration and share for United Bank & Trust.
- Increased 1997 commercial loan closings over 1996 through aggressive promotion and marketing; personally responsible for developing and managing capital investment programs for approximately 45 Cincinnati businesses.
- Effectively promoted visibility of United Bank & Trust through planning and presentation of seminars for clients and through participation at conferences and trade shows.

Assistant Vice President, Private Banking Group *(1993–95)*
Managed promotion and sales of complete breadth of financial products and services to portfolio of 60 clients characterized by high net worth. Responsibilities included origination, underwriting, and closing of commercial/consumer loans.

- Successfully restructured entire $3.2MM portfolio in noncompliance with bank loan policy.

1991–93 DIME SAVINGS BANK • Akron, OH
Commercial Banking Officer, Commercial Lending Division
Managed $8MM loan portfolio for savings bank with assets of $7 billion. Oversaw new business development, financial data collection, and analysis.

- Increased portfolio by 50% while maintaining delinquency ratio of 2%.
- Assisted in development of Departmental Strategic Plan.
- Regular guest speaker at regional forums throughout Ohio.

70

PROFESSIONAL EXPERIENCE (cont'd.)

1988–91   THE BANK OF CINCINNATI • Cincinnati, OH
            Lending Officer, Lending Group
            Elected to bank officer's position to assist in formation and
            establishment of commercial bank.
            • Managed consumer/mortgage lending department, generating
              $12MM in consumer loans and $6MM in commercial loans for
              fiscal year 1990.
            • Created and implemented correspondent mortgage lending
              program with extensive work in secondary market to liquidate
              bank's existing mortgage loan portfolio.

EDUCATION   HARVARD UNIVERSITY • Cambridge, MA
            B.A., Political Science (1987)
            • Foreign study included work at Richmond College in London and
              the Sorbonne in Paris.

            *Successfully completed continuing professional development programs
            in variety of disciplines including:*

            • RMA Cash Flow Analysis (1997)
            • Forum Corporation Dynamic Selling (1996)
            • RMA Uniform Credit Analysis (1994)

PROFESSIONAL AFFILIATIONS

• Akron Community College Foundation   Akron, OH
  Board Member, 1996–Present
• Cincinnati Business Outreach Center   Cincinnati, OH
  Loan Committee Member, 1995–Present
• Habitat for Humanity   Akron, OH
  Volunteer, 1995–Present
• Taking Care of Business   Cincinnati, OH
  Advisor/Member, 1994–Present
• Cincinnati Economic Development Corp.   Cincinnati, OH
  Board Member, 1994–Present

11—BANKING EXECUTIVE

Harry Ankenbaur
16 Oak Tree Road
Collinsville, Illinois 62234
(618) 555-4448

> Author's favorite. Excellent graphic display indicating past achievements.
>
> Resume is full of quantifiable accomplishments as indicated by the bullets.
>
> A lot of information in an easy-to-read format backed by impressive Civic Involvement.

Senior Banking Executive
Retail Lending / Collections / Property/Facility Management

Senior Operating Executive in the banking industry with proven leadership skills and expertise in refining credit quality, returns on investment and capital, cross-marketing, asset growth, and analysis of risk-adjusted returns on mortgage lending. Successfully orchestrated the successful centralization of lending, collections, underwriting, and loan processing procedures to maximize asset performance. Early background includes regional operations and training management for a St. Louis-based investment company, supervising offices in 12 Midwestern states.

Career Highlights

BANK OF ALTON, Alton, IL, 1993-Present

Senior Vice President/Lending Officer
Recruited to bolster the lending operation for this community bank established in 1956, now serving 25,000 customers with $170 million in total deposits.

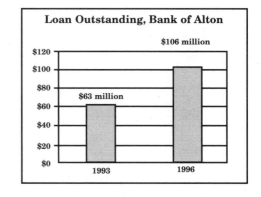

- Centralized the Collection and Secondary Mortgage Departments. Sold real estate loans in the Secondary Market totaling more than $20 million from 1993 to 1995. Collection of recovery accounts exceeded charge-offs for the last 3 years.
- Increased loan outstanding from $63 million to $106 million.
- Decreased classified loans from $10 million to $3 million.
- Implemented an Indirect Lending Department now serving more than 3,000 loan customers.

CENTRALBANK, Fairview Heights, IL, 1980-1993

[formerly known as Southern Illinois Bank (SIB), purchased by Central Bank System, Inc. (CBSI) in 1985 and sold to Firstbank of Illinois Company in 1991]

Senior Vice President/Retail Lender
Provided leadership for Indirect Lending, Direct Lending, Collections, Property Management and Facility Management departments. Supervised 42 employees. Following the 1991 merger of two United Illinois Bank branches, formulated centralization of all collections into a single department servicing more than 16,000 retail loan accounts with a total net outstanding of $289 million.

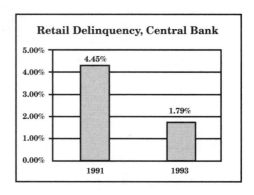

11—BANKING EXECUTIVE (*CONT.*)

CENTRALBANK, Fairview Heights, IL, 1980-1993
(cont.)

Supervised Property Management, including 11 facilities complete with a warehouse operation and a courier system operating between all locations.

Appointed Security Officer for the eleven locations in 1992. Established a written Security Manual conforming to the new Security Act and presented training sessions at all locations.

- Reduced retail delinquency from 4.45% in 1991 to 1.79% in 1993.
- Assumed additional responsibility of servicing all commercial collections deemed a Workout Status. Accepted the responsibility of managing the Bank's Other Retail Estate Owned (ORE). Reduced ORE from $1.1 million to $700,000.
- Centralized the Indirect Lending area into a single location servicing 46 auto dealers in the Metro-East area. With two underwriters and a clerical staff of four, processed more than 1,000 applications each month. Purchased $11 million in auto loan paper in the first three months of 1993 for a net gain in outstanding of $3 million.
- With the centralization of the collection department, reduced the amount of chargeoffs from 1.67% in 1991 to 0.85% in 1992.
- Implemented the Empirica Scoring System and a Quality Evaluation Guide to ensure quality asset growth and significantly reduce chargeoffs. Decreased 1993 1Q bankruptcy filings 50% compared to 1992 1Q.
- Centralized all Direct Lending into a single department processing approximately 200 direct loan applications per month. Provided a centralized approval process for Central Bank's 11 locations.

Vice President (1986)
Assistant Vice President (1980)
Successfully organized and set up a Secondary Mortgage Department for Central Bank in 1989. Originally started this effort with myself as the sole underwriter for FNMA. Formulated a staff that grew the department to $12 million in FNMA loans and serviced the direct real estate loans totaling $60 million. FNMA delinquency was zero when I left Central Bank.

Civic Involvement

- Illinois Center for Autism, 1989-Present; Chairman of the Board 1992-1993
- Salvation Army in Belleville, Board of Directors, 1989-Present
- Fairview Heights Rotary, Past President and holder of the Paul Harris Foundation Award; Proclaimed Rotarian of the Year in 1990
- Served as the District Treasurer for the Student Exchange Program in 1989
- Appointed to the Economic Development Commission in 1989
- Quest - Served on the Board of this Belleville Diocese Retreat Program
- Served on the Teens Encounter Christ (TEC) retreats for high school students for 10 years
- St. Henry's Parish - Finance Committee since 1991

Education

- Graduate of Illinois Bankers School of Southern Illinois University
- Applied Banking Diploma, American Institute of Banking
- Participated in Lending Management with Schesunoff
- Conference Leadership Training Course, American Investment Company

References Available Upon Request

12—BANK TELLER

CYNTHIA L. KOHL
h Mill Street • Nyack, New York 10960 • (914) 555-3160

OBJECTIVE
Seeking professional and personal growth in a position utilizing extensive *Banking/Customer Service* experience.

SUMMARY OF QUALIFICATIONS
A detail oriented banking professional with extensive experience in positions requiring superior interpersonal, communication and problem solving skills. Demonstrated ability to function effectively under all types of circumstances while maintaining a clear perspective of goals to be accomplished. Excellent leadership, time management and decision making abilities. Experienced in the use of computerized equipment to access, store and track pertinent information. Computer proficiency includes Windows 3.1, Microsoft Works and WordPerfect 6.1.

PROFESSIONAL EXPERIENCE
Manufacturers & Traders Trust - Nyack, New York
Bank Teller • 1994/Present
- Assist bank customers in processing all types of checking and savings transactions.
- Provide written materials and forms to aid in coordinating additional banking functions.
- Interface with customers to research account information and resolve issues.
- Supervise three Tellers in the absence of Head Teller.
- Utilize various types of computerized equipment/software to process information and verify accuracy.

Selected Accomplishments:
- *Received numerous quarterly performance awards for consistent levels of superior service.*

Food Emporium (A&P Corporation) Greenwich, Connecticut
Co-Manager • 1993/1994
- Managed day-to-day front end operations of this $12 million upscale supermarket.
- Coordinated customer assistance through detailed knowledge of each department.
- Monitored cashier performance and oversaw training of all new employees.
- Participated in interviewing and hiring of all front-end personnel.
- Maintained accurate pricing, inventory tracking and ordering to maximize productivity.
- Created schedules to ensure sufficient personnel during peak hours and submitted payroll documentation.

Edwards Super Food Store (formerly Finast Supermarket Corp.) - Amityville, New York
Assistant Front End Manager/Department Manager • 1983/1993
- Worked full and part time during school, fulfilling progressively responsible duties.
- Processed returns, verified deliveries and provided support for Front End Manager.
- Scheduled cashier assignments to ensure adequate coverage during peak hours, monitored performance and offered recommendations to increase productivity.

EDUCATION
- **State University of New York at Stony Brook**- Stony Brook, New York
  Bachelor of Arts • Social Science Interdisciplinary

13—BOOKKEEPER

MARY-LINDA WILLIAMS
TELEPHONE/VOICE MAIL: 953-555-4334
2657 WANGELWOOD TRAILS
MOBILE: 953-555-7997
JEFFERSON, TEXAS 77903

> *Strong resume with keywords to open the document. This candidate does not have formal accounting education so it is left off the document; her experience is the selling point in this resume.*

Bookkeeper/Auditor

Career professional with extensive background in accounting support and office management functions. Detail-oriented; streamline procedures; cut administrative operations cost. Bring analytical and organizational skills combined with core competencies in...

Accounting Support:		**Information Technologies:**	
Accounts Receivable	Payroll	IBM	MS Word
Accounts Payable	Financial Reporting	Windows 95	Macintosh
ProfessionalWrite	Tax Documentation		
Reconciliations			
Auditing			

...to contribute to the productivity and profitability of your organization.

Professional Experience

HOLIDAY INN
Jefferson, Texas
BOOKKEEPER/AUDITOR   1993 TO PRESENT

Sole responsibility for bookkeeping/auditing function for three business entities.
- **Set up and directed bookkeeping/auditing function for hotel, restaurant and club.**
- **Originated all regulatory and accounting procedures for club in compliance with state Liquor Control Board codes.**
- **Set up accounting forms for hotel, restaurant and club.**
- **Train new and relief auditors.**
- **Authored training manual that became the company standard.**

HARTE-HANKS COMMUNITY NEWSPAPERS
Tyler, Texas
PAYROLL ACCOUNTING CLERK   1991 TO 1993

Accountable to Financial Director for time sheet auditing and payroll processing including reconciliation of payroll accounts. Purchased office and printing supplies, balanced daily cash receipts, prepared daily bank deposits and reconciled bank accounts for nine community newspapers.

- **Contributed to special projects for Financial Director, CFO and Publisher.**
- **Achieved 10% cost reduction in supply purchasing by narrowing vendors to "preferred only" status.**

CITY AIR CONDITIONING COMPANY
Tyler, Texas
OFFICE MANAGER   1988 TO 1991

Charged with full decision-making authority during owner's absence. Directed administrative staff of four and technical staff of six. Full responsibility for payroll, purchasing and account audit.
- **Functioned as Estimator; authorized to conduct project bids.**
- **Expedited difficult collections.**

Professional Associations:

THE AMERICAN INSTITUTE OF PROFESSIONAL BOOKKEEPERS

Alexandra Lynn Zolac ▣ Television and Video Production

▣ Professional Profile

Produce daily interview segments airing on Bloomberg Terminals, Bloomberg News and BIT (Bloomberg Information Television) world-wide. Handle live, two-way satellite and remote pre-production, production, post-production, and editing.

Set-up and tape live news programming and studio, satellite and remote television interviews with business leaders, heads-of-state, celebrities and newsmakers.

Professionally committed, focused, responsible, and cool in chaos. Utilize an effective combination of technical abilities, creativity, and good humor to produce excellent interviews and programming in normal or unforeseen situations.

▣ Technical Equipment and Abilities

Switchers	▫ Abekas A34, PDG 418 Videotek
Cameras	▫ Sony BetaCam, Hitachi studio camera
Remote Controls	▫ Fujinon, Radamec advanced robotic cameras
Audio	▫ Mackie 32x8 Audio Console
Routers	▫ BTS Digital CP-300, SYC-3200 Sigma, Codec
Editing	▫ AVID / Video cube, Abekas A34, RM-450, PVE-500
Computers	▫ Audio Video Capture Computer (AVC), PC, Mac

▫ Pre-and-post production and basic editing for live and remote newscasts.
▫ Live production transmission utilizing multi-media computer.
▫ Dubbing and duplication of tapes and tape format changes.
▫ On-air switches and master control duties. Live daily newscast teleprompting.
▫ Remote and studio control cameras. Live and taped.

▣ Employment and Education

Bloomberg, LP, New York, NY Video Forum Technician and Segment Director	**1996 to present**
WLNY, Melville, NY Studio Technician and Editor for live daily news show	**1994 to 1996**
Bachelor of Fine Arts in Broadcasting New York Institute of Technology, NY	**1994**

610 Broadway, Amityville, NY 11701   ▫   516-555-5555   ▫   alexz@aol.com

15—BUSINESS CONSULTANT

JOHN F. NEYE
750 Northwest 129th Street
Margate, FL 33063

Notice the focus is on his accomplishments and Areas of Expertise, not his actual employment.

E-Mail: Neye21@msn.com
Mobile: (954) 555-3855
Home: (954) 555-8058

INDEPENDENT MANAGEMENT CONSULTANT / SUCCESS COACH
Driving Organizational Change, Process Redesign, Revenue & Profit Enhancement
* Start-Up Opportunities / Turn Around Situations / Quality Improvement

---

Dynamic, top-performing hands-on management consultant with 20 years professional experience. Expert in organizational assessment, strategic planning/implementation, technology utilization, and process development. Pioneer in the design and delivery of innovative, bottom-line change management programs that have generated **millions of dollars in new revenues/cost-savings** through redesign/restructuring of internal operating production and business processes - consistent with short/long term organizational objectives.

Experienced in identifying and capitalizing upon market opportunities to introduce new products/services, reposition existing ones, and **drive revenue/earnings growth in highly competitive markets**. Visionary leadership in turning around under-performing operations and start-up opportunities through team leadership, building key alliances, and implementing quality control management systems.

AREAS OF EXPERTISE

* Finance management - P&L accountability
* Long/short term strategic planning and execution
* Sales/marketing/business development
* Organization and time management
* Cross functional training and leadership
* Total quality management

* Behavior modification - change management
* Complex project planning and management
* Productivity and performance enhancement
* Policy and procedure development and redesign
* Multiple location management and logistics
* Delivering top-rated customer service

HIGHLIGHTS OF EMPLOYMENT

TIRE KINGDOM, INC., West Palm Beach, FL		1985 - Present
Vice President of Special Projects / Director of Service Sales	**(1996 - Present)**	
Vice President of Customer Service	**(1994 - 1995)**	
Vice President of Southeast Region	**(1988 - 1994)**	
District Manager / Store manger	**(1985 - 1988)**	
C. WAYNE MOTOR COMPANY, Plattsburgh, NY		1983
Service Manager		
COMMERCIAL & INDUSTRIAL REAL ESTATE, Plattsburgh, NY		1980 - 1983
Business Development		
SAFETY STEERING & TIRE COMPANY, Plattsburgh, NY		1973 - 1979
Owner / Operator		
PERU CENTRAL HIGH SCHOOL, Peru, NY		1967 - 1972
Physical Education and Health Teacher		
Varsity Basketball & Football Coach		

EDUCATION & TRAINING

ITHACA COLLEGE, Ithaca, NY
Bachelor of Science Degree, 1967

- References and Supporting Documentation Furnished Upon Request -

16—CALL CENTER MANAGER

Evelyn Arbuckle

Call Center Manager...Banking Operations Supervisor

- More than five years of lending experience in the banking industry, most recently as the managing director of Roosevelt Bank's first and highly successful Loans-By-Phone call center.

- Proven ability to build a dynamic sales culture with high standards for in-depth product knowledge, continuous training, and quality customer service.

- Key resource for compliance and staffing issues, communications systems, and special programs.

Professional Experience

ROOSEVELT BANK, St. Louis, Missouri   1994 to Present
Call Center Manager, Loans-By-Phone *(1996-Present)*

Promoted to oversee the start-up and operation of a new consumer and mortgage loan call center accepting applications from Missouri, Kansas, and Southern Illinois, expected to generate $2 million per month in closed loan volume. Direct, train, schedule, supervise, and motivate a sales staff of six part-time and full-time employees responsible for selling a wide range of banking products.

Track and analyze an average of 150 incoming calls per day to ensure 12-hour phone coverage during slack and peak periods. Prepare weekly and monthly reports for senior management. Train sales staff to maintain up-to-date product knowledge and cross-selling opportunities.

Exceeded all first-year department performance goals:
- Generated more than $46 million in closed loan volume (nearly 200% of forecast) in 1996, representing about $4 million per month ($1 million per employee).
- Helped position Roosevelt as the third largest home equity lender in St. Louis with $53 million in home equity loans closed. Accounting for 82% of the bank's total home equity loan production, the Loan-By-Phone department itself would have ranked 4th among the area's Top 30 lending institutions.
- Managed three sales people whose individual production averaged higher than three of the Top 30 lenders. The leading sales producer would have ranked 20th on the Top 30 list for 1996.
- Personally closed more than $6 million in loans while concurrently managing department.

Additional experience:
Loan Originator *(1994-1996):* Managed four branches originating $10 million in mortgage and consumer loans per year. Consistently ranked as a top producer.

Previous background includes real estate sales, small business management, training, and three years as a Loan Officer with a savings and loan.

Education/Training

Bachelor of Science Degree, Notre Dame College, St. Louis, Missouri
Licensed Realtor, State of Missouri

References Available Upon Request

17—CASE MANAGER

SAMANTHA WILSON
151 W. Passaic Street, Rochelle Park, New Jersey 07662
Tel: (201) 555-3772 Email: xxxxxxx@aol.com

CASE MANAGER

counseling   crisis intervention teaching
advocacy   documentation

PROFESSIONAL SUMMARY

Strong practical and theoretical foundation in various therapeutic and intervention models.
Excellent skills in assessing client's needs with commitment to enhancing personal growth.
Member of interdisciplinary team in the development and implementation of treatment plans.
Supportive and sensitive with the demonstrated ability to teach and guide others; recognized for inspiring the confidence and trust of clients.
Honored as Housing Counselor of the Year at West Bergen Mental Healthcare.

PROFESSIONAL EXPERIENCE

WEST BERGEN MENTAL HEALTHCARE, Ridgewood, New Jersey   1991-Present

CASE MANAGER

Provide and manage comprehensive direct services to the chronically mentally ill population of a mental health organization serving 13 towns within northern Bergen County, New Jersey:
  Utilize excellent clinical skills in screening clients' psychiatric status, providing supportive counseling and in crisis intervention/de-escalation.
  Monitor clients' medical conditions and self-administration of medications; coordinate, relay and document both psychiatric and medical information/data.
  Guide clients in enhancing self-sufficiency and building self-confidence level by teaching daily living skills.
  Prepare thorough and accurate documentation of each client's overall condition on daily basis.
  Collaborate with peers on clinical case issues and complete applications for allocation of community resources for clients.
  Preside over client community meetings.
  Communicate/collaborate effectively with others as member of interdisciplinary treatment team and participate in weekly staff meetings.

ACHIEVEMENTS

Conducted presentations to community groups.
Appeared on community access channels for 5 New Jersey cable networks.
Expanded referral network through aggressive marketing to state and community officials.
Designed marketing material and educational curriculum.

EDUCATION

B.A. in Psychology
William Paterson College, Wayne, New Jersey

PROFESSIONAL DEVELOPMENT

Company sponsored training includes:
Attending Annual Schizophrenia Conference, Columbia University
Numerous seminars through the American Healthcare Institute

DAWN M. PITERA

151 W. Passaic Street • Rochelle Park, New Jersey 07662 • **(201) 555 -3772**

CORPORATE / EXECUTIVE CHEF

An accomplished and extensive culinary and management career directing exclusive high volume restaurant, hotel and catering operations worldwide. Strong leadership and management qualifications combine with outstanding cross-cultural communications, interpersonal and team building skills. A significant contributor to cost reductions and revenue/profit growth through productivity, operational efficiency and service/quality improvements. Multilingual proficiency in French, German and English; conversant in Spanish. Demonstrated expertise in:

- **Food/Beverage/Labor Cost Controls**
- **Staff Training & Development**
- **Purchasing & Inventory Management**
- **Innovative Menu Development & Planning**
- **Quality Assurance & Control**

- **New Facilities Start-Up**
- **Budget Administration**
- **Food Preparation & Presentation**
- **Special Events Management**
- **Customer Service & Guest Relations**

PROFESSIONAL EXPERIENCE

UPPER MONTCLAIR COUNTRY CLUB • Clifton, New Jersey

EXECUTIVE CHEF • 1993 - Present

Coordinate and manage food service and catering operations for a privately-owned golf club with 950 members and host of annual NFL/Cadillac Tournament. Direct staff of 30 from front to back of house in all aspects of restaurant and catering operations including menu planning, recipe development, food preparation and events management. Control equipment, food and beverage purchasing, inventory, receiving and food/labor costs within operating budget guidelines. Develop and train all cooks, kitchen and wait staffs.

Accomplishments:

- ◆ *Grew sales from $1.5 million to $2.5 million annually.*
- ◆ *Cut food costs 10%, reduced labor costs and eliminated overtime by utilizing state-of-the-art equipment.*
- ◆ *Strengthened communications and cooperation between departments.*
- ◆ *Enhanced food preparation efficiency by providing comprehensive staff training.*
- ◆ *Instituted food protection/sanitation program for all cooks and food preparation staff.*

LOEWE/MARRIOTT GLEN POINTE HOTEL • Teaneck, New Jersey

SPECIALTY EXECUTIVE CHEF • 1991 - 1993

Directed and trained staff of 50 in food preparation/presentation and specialty items for hotel, restaurant and catering operation with annual sales of $5.5 million. Managed all preparations for on- and off-premise special events for up to 3000 hosted guests. Planned and designed menus. Coordinated all purchasing and related functions.

Accomplishments:

- ◆ *Cut food costs 5% by training cooks on proper equipment handling.*
- ◆ *Significantly upgraded quality and efficiency of food operations and service standards by implementing comprehensive staff training and food protection certification programs.*
- ◆ *Boosted sales through introduction of food and wine tasting events.*

DAWN M. PITERA   (201) 555-3772   Page Two

PROFESSIONAL EXPERIENCE continued...

LE PAVILLION CORPORATION - PRUNELLE • New York, New York

CHEF / MANAGING DIRECTOR • 1983 - 1991

Created and managed popular French restaurant attracting many famous personalities; recognized as "one of the best restaurants in New York." Directed all food preparation / presentation, purchasing activities, menu development / planning, and events management. Managed entire staff of 40 employees, including recruitment and training. Ensured adherence to quality of food operations and service standards.

Accomplishments:

♦ *Built sales to $3.6 million in 3 years maintaining food costs at 28-30% with net profit of 20%.*
♦ *Elevated restaurant from 1-star to 4-star rating by 3rd year and designed first class menus.*
♦ *Frequently invited as guest chef on radio/television shows on all 3 major networks and cable.*
♦ *Developed and presented programs demonstrating cooking techniques to U.N. ambassadors' wives from around the world.*
♦ *Participated in philanthropic functions for March of Dimes and various community organizations.*

LA GAULOISE RESTAURANT • New York, New York

EXECUTIVE CHEF • 1979 - 1983

Managed purchasing, inventory control, shipping/receiving and personnel functions including hiring, training and scheduling staff for $1.2 million French bistro. Innovated recipes and directed menu planning, food preparation and events management.

PREVIOUS PROFESSIONAL EXPERIENCE at exclusive hotels and restaurants including *Park Hotel Krefelder Hof* (Krefeld, West Germany), *Tiffany's* (Montreal, Canada), *Sporthotel Mohnenfluh* (Schröcken, Austria), *Le Vieux Chalet,* (La Clusaz, France) *Cercle National Des Armees* (Paris, France) and *Hotel/Restaurant Du Haut-Koenigbourg* (Alsace, France).

EDUCATION

Certificate of Professional Aptitude
CULINARY ACADEMY • Strasbourg, France

Graduate - Hotel and Restaurant Management
ECOLE HOTELIERE ET DE TOURISME • Strasbourg, France

HONORS

Recognized as 3rd out of 72 professional chefs in the Alsace, France region
Named among *"New York's Best Young Chefs"* by Moira Hodgson of New York Times in 1980
Named "rising star" by Gael Greene of *New Yorker* magazine in 1979

PROFESSIONAL AFFILIATIONS

Jury Member - French Culinary Institute of New York
Honorary President - Alsace Chefs of North America
Member - Society Culinaire Philanthropique

19—CIVIL ENGINEER

JAKE D. BUILDER, P.E.

1234 Cherry Lane, Canton, Georgia 30014 • (770) 555-4567

SENIOR OPERATING ENGINEER • PROJECT MANAGER
DIRECTOR OF PUBLIC WORKS • DIRECTOR OF ENGINEERING

Distinguished Senior Management Executive with considerable, diverse experience in engineering, operations, and project management. Specific expertise is focused on public utilities. Registered Professional Engineer #GPE000X, State of Georgia. Skilled in:

- Problem Solving
- Budget Administration
- Computer Systems
- Supervision & Training
- Engineering & Project Management

- Operations & Financial Management
- Organizational Administration
- Quality & Productivity Improvement
- Staff Coordination & Development
- Strategic Planning & Forecasting

PROFESSIONAL EXPERIENCE

CITY OF CANTON, Georgia - 1994 to present

DIRECTOR OF PUBLIC WORKS reporting to the City Manager. Oversee administration of a $24 million annual Operating Budget and a Capital Improvement Budget of $4.5 million. Functional authority includes direct supervision of the City Engineer, Traffic Engineer, Public Works Superintendent, Integrated Waste Manager, Utilities and Facilities Maintenance Manager, and 90+ employees.

Scope of responsibility includes directing all Public Works Department activities.. Involved with administering the Capital Improvement Program; Engineering Design; Construction Inspection; Flood Plain Administration; Mapping; Transportation Planning; Traffic Operation; Street, Parks, Parkway, Storm Drain, City Facilities, Grounds, Fleet and Equipment Maintenance; City Water and Sewer Systems; Recycling Programs; Refuse Collection and Street Sweeping.

- Developed the city's first Capital Improvement Plan including buildout systems funded by development impact fees and a five year improvement plan.

- Negotiated a multiagency funding agreement facilitating completion and construction of the new Riverside Parkway.

- Implemented an automated refuse service system increasing employee efficiency, reducing costs and workmen's compensation claims.

- Developed and administered a $6 million Water Bond Certificate of Participation for construction of master planned water facilities.

Résumé	JAKE D. BUILDER, P. E.	*Page Two*

PROFESSIONAL EXPERIENCE (*Continued*)

BAKERSVILLE, Mississippi - 1985 to 1994

DIRECTOR OF ENGINEERING reporting to the County Administrator. Oversaw and administered annual Operational Budget of $4.9+ million and a Capital Improvement Program budget of $3.6+ million. Operational responsibilities included direct supervision of 4 Development, Construction, Right-of-Way, and Traffic Engineers and 48 employees.

Coordinated operational activities including: Engineering Design, Construction Inspection, Construction and Legal Surveying, Right-of-Way Acquisition, Development Plan Review, Traffic Operations, Transportation Planning, Pavement Management, Drainage Plan Review, Flood Plain Administration, Construction Permitting, and Construction and Engineering Services Contract Administration.

- Developed and implemented a five-year Capital Improvement Program which required solicitation of outside funding sources for project implementation.

- Negotiated and implemented intergovernmental agreements for planning and construction projects approved by the Mississippi Department of Highways, Urban Drainage and Flood Control District.

- Revised and administered flood plain regulations to maintain eligibility in the Federal Emergency Management Agency (FEMA) flood insurance program.

MEMBERSHIPS

- American Public Works Association (APWA), Kansas City, Missouri
- American Water Works Association (AWWA), Denver, Colorado
- Fairbanks County Council of Governments Technical, Fairbanks, Alaska

ACCOMPLISHMENTS AND AWARDS

- *Outstanding Engineer's Performance Award*, County Administrator, Fairbanks, Alaska, 1996
- *National Association of Counties Achievement Award,* Denver, Colorado, 1992
- *Schuette Management Award*, Schuette and Associates, Fallbrook, California, 1987

EDUCATION

University of San Diego, San Diego, California - 30 credit hours
    Graduate Studies in Business Administration

University of Georgia, Marietta, Georgia
    Bachelor of Science in Civil Engineering

Appropriate personal and professional references are available

LISA JOHNSON

402 Rally Drive • Belleville, Illinois 62221 • (618) 555

SUMMARY OF QUALIFICATIONS

Six years experience as the Head Coach of highly competitive basketball, volleyball, and softball teams for a quality NAIA women's intercollegiate athletic program in the American Midwest Conference (formerly the Show-Me Collegiate Conference). A successful track record for graduating student-athletes (97%), managing athletic programs, recruiting, and developing talent at the high school, junior varsity, and varsity levels. Actively involved in faculty, athletic, and student affairs committees. Proven athletic administration skills include:

Budgeting	Travel Planning	Public Relations
Staffing/Scheduling	Academic Standards	Recreational Programs

- Graduated 60 Academic All Conference volleyball and softball players in the last five years. Coached volleyball teams that qualified for post-season play each year and softball teams that have finished no less than fourth in the conference, including a regional championship and a 4th place national finish.
- Member of the American Volleyball Coaches Association, National Softball Coaches Association, and National Association of Intercollegiate Athletics.
- Served as the Director of Public Relations for the St. Louis Steamers Soccer Club from 1985-1987. Accustomed to developing a strong rapport with local media contacts.
- Hold an M.A. Degree in Speech Communication from Southern Illinois University at Edwardsville and a B.A. Degree in Communication from St. Louis University. Earned Academic All Conference honors in basketball as an undergraduate. Taught various Physical Education (activities and theory) and Communication courses over the last six years including Public Speaking, Interpersonal Skills, Small Groups, and Persuasion.

COACHING HIGHLIGHTS/COMMITTEE WORK

Head Volleyball/Softball Coach, McKauliffe College, Livingston, Illinois   1989-Present

- 1995: American Midwest Rating Committee, Volleyball
- 1995: 4th Place, NAIA National Softball Championship, the highest national tournament finish for any athletic team at McKauliffe College
- 1995: NSCA Exposure Camp Director
- 1995: NAIA Executive Committee, Softball
- 1995: NAIA Midwest Region Champions, Softball
- 1995: NAIA Midwest Region Coach of the Year, Softball
- 1994: NSCA Exposure Camp Staff
- 1993: NAIA District #20 Chair, Softball
- 1992: Show-Me Collegiate Conference Champions, Volleyball
- 1992: Show-Me Collegiate Conference Coach of the Year, Volleyball

	1990	1991	1992	1993	1994	1995	Totals
SOFTBALL							
Won/Loss	29-14	19-9	21-10	22-8	28-18	41-16	160-75 (68%)
NAIA All Americans	1	n/a	n/a	n/a	n/a	2	3
All Conference	2	3	3	4	5	5	22
Academic All Conference	1	1	5	7	8	9	31
VOLLEYBALL							
Won/Loss	18-10	24-20	18-11	25-13	28-13	16-21	129-88 (60%)
NAIA All Americans	n/a	n/a	1	1	2	2	6
All Conference	2	2	3	4	3	3	17
Academic All Conference	2	2	5	6	7	7	29

An author's favorite—a creative and fun resume.

Excellent graphic presentation at the bottom.

Solid Summary of Qualifications section.

21—COMPUTER PROGRAMMER

Good use of keywords in the front end of the resume.
Very qualifications-oriented, almost like a CV.
Simple, easy to read, and very focused on skills.

Frank Benuscak

15 North Mill Street - Nyack, New York 10960 - (914) 555-3160

Profile

A results-oriented **COMPUTER PROGRAMMER** with extensive experience in all stages of design, coding and testing for multi-user systems. Background encompasses extensive experience in software analysis and development in demanding hi-tech environments. Demonstrated ability in providing end-user training and developing on-line Help documentation and user manuals.

Software

Assembly (16 Bit) - Liberty Basic - Visual Basic 4.0, 5.0 - VBA - MS TSQL - MS Access - MS SQL Server 6.5 - Crystal Reports - C - C++ - DOS - UNIX - FORTRAN - COBOL - Microsoft Office

Hardware/OS

IBM -PC's and Compatibles -Windows NT, 95,& 3.1

Professional Experience

Institute of Energetic Researches of Academy of Sciences - Washington, DC - 1990/Present
Programmer/Analyst - 1995/Present
Directed all stages of design, coding and testing of the Environmental Management System (EMS) project concerning the ecological safety of power plants.
* Designed multiple forms allowing employees to enter research resulting and forecasting information into the EMS Database.
* Developed on-line Help documentation and user manuals as a result of gathering information from client interaction.

Programmer - 1990/1995
Participated in the development and implementation of the Coolant Flow Regimes Calculating System (CFRCS) for the needs of the physics and hydrodynamics laboratories of the power plants.
* Utilized numerous scientific formulas and created Assembly Language (16 bit) with later transition to Liberty Basic for Windows.

Scientific Research Center for the Atomic Power Engineering - Moscow, Russia - 1987/1990
Engineer

Scientific Research Institute for the Nuclear Power Stations - Moscow, Russia - 1979/1987
Engineer

Education

NRI Schools - McGraw - Hill Companies -Washington, DC
Programming Training

Moscow Physics Engineering University - Moscow, Russia
Bachelor of Science - Physics and Engineering

22—CUSTOMER SERVICE

LAUREN A. JOHNSON

15 North Mill Street • Nyack, New York 10960 • (914) 555-3160

CUSTOMER SERVICE/SALES PROFESSIONAL

PROFILE

A results oriented clients services professional with extensive experience in client relations and marketing. Excellent sales skills include thorough product knowledge and the ability to convey all pertinent facts to customers, staff and management. Demonstrated ability in assessing problem areas and offering recommendations resulting in increases in productivity and profitability. Background encompasses the ability to establish and build positive, solid relationships with clients and all levels of management. Computer experience includes Windows 95, Microsoft Office and WordPerfect.

AREAS OF EXPERTISE

Client Services • Sales/Marketing Troubleshooting • Market Penetration • New Product Launches
Account Retention • Cold Calling • Vendor Relations • Product Knowledge

PROFESSIONAL EXPERIENCE

Ferring Pharmaceuticals Inc. - Tarrytown, New York • 1993/Present
Customer Service Representative /Product Specialist
Generated sales by prospecting qualified leads and following through on customer inquires. Maintained records and tracked sales progress on specially designed industry software. Assisted existing and prospective customers by detailing services, products and drug therapies. Conducted in-service training for medical staffs on the administration of specific intravenous medication and provided technical support for physicians, nurses and pharmacists. Attended pharmaceutical conventions throughout the United States establishing sales contacts throughout the industry and handled follow-up.
Selected Accomplishments:
- *Received promotion after reorganization of sales department.*
- *Developed training/customer service manuals for specific product line.*
- *Selected to train outside contracted sales force on intravenous, infertility drug therapy.*
- *Received awards for consistently achieving highest telephone usage time for client services.*

March of Dimes Birth Defects Foundation - White Plains, New York • 1993
Community Director
Developed and delivered presentations for this non-for-profit organization to schools, community groups and health care facilities focusing on women and children's health issues. Procured donations from Fortune 500 corporations and individual businesses for various fund raising events. Worked with several local municipalities to provide prenatal care to economically disadvantaged women.
Selected Accomplishments:
- *Generated over $50,000 in revenue in the* <u>WalkAmerica</u> *"Campaign for Healthier Babies".*

Stanley H. Kaplan Test Preparation Centers -Nanuet, New York • 1992/1993
Sales/Marketing Associate
Participated in the development of marketing and sales programs for high school, college and graduate level test preparation courses for this subsidary of <u>The Washington Post</u>. Fulfilled various administrative functions, trained and supervised staff regarding operating policies. Acted as resource of information and ensured accurate scheduling of classes to optimize instructor assignments. Assumed additional management responsibilities as needed in absence of senior managers.

Teacher
Implemented and conducted lesson plans for high school students taking the PSAT and SAT. Provided individual tutoring for students.

EDUCATION

- **Dominican College** - Orangeburg, New York
  Accelerated Bachelor of Science Nursing Program • 1993
- **Ramapo College** - Mahwah, New Jersey
  Bachelor of Arts • *Psychology & Literature* • *Cum Laude* • 1992

STELLA WILLIAMS
453 PEMBERTON CIRCLE, MIDDLETOWN, OH 12345

CUSTOMER SERVICE SUPERVISOR / MANAGER / TRAINER
20+ Years Success In Training and Mentoring Top Performance For Local and National Organizations
B.A. Degree / Led Numerous National Training Sessions

- Led seminars in strategies venting customer frustration at national and regional levels. Authored training implemented facility-wide for a national claims and correspondence center.
- Proven ability to build service-oriented teams who exceed expectations. Lead by example. Nominated #1 Associate by staff. Received departmental award for service excellence.
- Earned fast track promotions with a leading insurance company serving an older clientele. Managed a staff of 16 within just 4 years based on exceptional track record, top performance reviews, positive personal reviews and thank-yous from customers.
- High energy level, dynamic and proactive in handling customer complaints. Seasoned in answering inquiries over the phone and in writing - - understand when the situation warrants either means of response. Skilled in reviewing responses to pinpoint areas for quality improvement.
- Strong believer in and proponent of community activities - - sought numerous opportunities "giving back to the community" throughout career.

MANAGEMENT / CUSTOMER SERVICE EXPERIENCE

A MAJOR INSURANCE COMPANY, CINCINNATI, OH   **1985 to Present**
Top Performer - Peak Results - Customer Service - National Phone/Correspondence Center

ACTING SUPERVISOR (1994 to Present)
Standardized Processes, Earned Special Recognition, Empowered Staff of 32
- Took over a staff of 32 claims service representatives who handled telephone interaction with and authored correspondence to clients nationwide.
- Established goals and monitored conversations to assure service commitments were met and time was productively invested. Standardized and streamlined documentation, raising productivity 20 percent.
- Recruited, hired, trained and assessed individual performance and made recommendations for promotions based on merit.
- Counseled staff 1:1 as needed to motivate them and keep them at peak performance levels. Successfully mentored a 35 percent increase in claims handling within first 3 months.
- Identified a need to empower associates and made staff more autonomous by requiring completion of an Associate Certification Program, the first of its kind in company's history.
- Eliminated duplication of efforts by integrating a Windows-based centralized claims processing system that automatically updated claims representatives' activity.
- Nominated Associate of the Year by staff in 1995 based on dedication, reliability and commitment.

ASSISTANT CLAIMS INFORMATION SUPERVISOR (1990 to 1994)
Gave National Presentation, Developed Regional Training Session, Supervised Staff Of 16
- Oversaw 16 claims service representatives and coached an award-winning team - - staff received an award for excellence based on a 90 percent customer satisfaction rate.
- Asked to deliver a presentation to 500 key corporate officers and company managers on problem solving strategies.
- Developed and facilitated a regional training session for 200+ participants in proper client interaction entitled "Know When To Say When." Targeted specific ways to discuss claim and payment issues, held mock demonstrations, discussed best strategies for explaining coverage and venting customer frustrations.
- Opened and closed special lines for peak periods. Monitored phone systems, transferred representatives between queues, adjusted coverage to accommodate up to 5,000 incoming calls daily.

STELLA WILLIAMS
CUSTOMER SERVICE SUPERVISOR / MANAGER / TRAINER
2 of 2

MANAGEMENT / CUSTOMER SERVICE EXPERIENCE

A MAJOR INSURANCE COMPANY (cont.)

ASSISTANT SUPERVISOR - HYBRID CUSTOMER COMMUNICATION DIVISION (1989)
Systematized Quality Reviews Of Written Customer Communication
- Scrutinized the correspondence of 16 representatives who fielded incoming calls and responded to inquiries in writing.
- Earned special recognition from upper management for establishing standards for written communication and a systematic approach to quality reviews, a practice which previously did not exist.

CUSTOMER SERVICE REPRESENTATIVE - HYBRID CUSTOMER COMMUNICATION DIVISION (1988)
Met Department's Changing Needs With Cross-Training and Added Coverage
- Cross-trained customer communication and customer service representatives to effectively handle letter writing and direct telephone interaction. Taught both angles of services on a regular basis.
- Coordinated additional coverage to address special issues. Created and instructed a claims information unit to handle enrollment questions during a re-rate period. Handled all training and project supervision.

CUSTOMER SERVICE REPRESENTATIVE (1985 to 1987)
CUSTOMER SERVICE DIVISION (1987) / CUSTOMER COMMUNICATION DIVISION (1985 to 1987)
Trained New Hires, First Representative To Work In Both Divisions
- Learned and mastered various plans, enrollment and payment procedures, both on the phone (Customer Communication) and via correspondence (Customer Service). Wrote personal responses to incoming letters and interacted with customers directly on the phones.
- Frequently asked to train new hires. Successfully communicated the fundamentals of outstanding client relations, proper time management and personal interaction for maximum results.

TELLER ... OHIO FEDERAL SAVINGS & LOAN, CLEVELAND, OH   **1984 to 1985**

ACCOUNTING MANAGER ... WE SELL IT REAL ESTATE, CLEVELAND, OH   **1972 to 1984**
Accounting - Banking - Supervision - Insurance/Real Estate Transactions - Independent Realtor
- Promoted to Accounting Manager within 2 years. Supervised a staff of 3 processing monthly rental receipts for 1,120 units located in a 3-county area.
- Managed a $24 Million annual account base. Handled property/casualty, fire and renter's insurance sales and ongoing service - - cut collection time from net 42 days to net 31 through a more aggressive in-house strategy.

EDUCATION

INDIANA UNIVERSITY OF PENNSYLVANIA, INDIANA, PA ... B.A. in Liberal Arts

CONTINUING EDUCATION ... Attended seminars in human relations, customer service, recruiting, training, claims processing. Completed a course in Windows 95 and internet applications.

PROFESSIONAL AND COMMUNITY AFFILIATIONS

Instructor - "Know When To Say When In Service," "Diffusing the Angry Customer," "Developmental Service Concepts." Feedback Coordinator. Tour Guide. Member and Co-Founder, "LOVE" Team (Lunchtime Outreach Visits To The Elderly). Tutor for Special Needs Children. Special Olympics Volunteer.

SHIRLEY SMILEY

ST TOWN, PA   **555.555.4321**

DENTAL HYGIENIST / VOCATIONAL INSTRUCTOR

Private General and Periodontal Practices, Streamlined Operations for Peak Efficiency
Self Starter Who Takes Initiative / Established Co-Op Program / Coached Award Winning Performances

DEMONSTRATED STRENGTHS

• Patient relations • Enthusiasm • Patient education • Training personnel • Accuracy in diagnostic testing •
• Productivity-oriented • Created and implemented time and cost saving ideas •

DENTAL HYGIENE EXPERIENCE

DENTAL HYGIENIST (Present Employer's Name Withheld for Confidentiality)   **1990 to Present**
High Visibility Position - 1 Location - General Dentistry - Cleaning, Diagnostic Testing And Assisting - Volunteer
Regularly To Handle Special Assignments

DENTAL HYGIENIST … DR. PETER ANDOLO, BROWNSTOWN, PA   **1978 to 1990**
Promoted To Full-Time In 1980 - 1 Location - General Dentistry - Cleaning, Diagnostic Testing And Assisting
With Procedures - Revamped Recall Procedures - Earned Special Recognition For "Going The Extra Mile"

- Consistently met a 45-minute turnaround at all positions without sacrificing quality of care or personal attention to patients.
- Establish and maintain an extraordinary rapport with patients. Thoroughly explain procedures to decrease their anxiety. Extensive focus on patient education, pinpoint areas for improvement and offer strategies to be combined with daily oral hygiene for optimal results.
- Performed bite wing and panoramic radiographs (exposing and developing), pit and fissure sealants, temporary filling placement, crown recementing and sterilization since 1978.
- Take the initiative to perform a "courtesy exam" prior to the doctor's exam. To date, assessments have been accurate.
- Entrusted as front line in toothache emergencies - - employ percussion testing and periapical film to identify source of discomfort.
- Came in while office was closed for the weekend to perform a prophylaxis and fluoride treatment prior to a patient's initial radiation therapy for a lymphoma of the left mandible. Received special recognition for exceptional work.
- Revamped patient recall procedures for a general practitioner. Implemented a tracking system that eliminated multiple mailings to patients' homes, slashed mailing expenses and cut preparation time.

INSTRUCTIONAL EXPERIENCE

INSTRUCTOR … VOCATIONAL CAREER & TECHNOLOGY CENTER, WINDBER, PA   **1971 to 1978**
Dental Assisting - Implemented Co-Op Program - Coached 2 Award-Winning Students

- Instructed high school students for a regional vocational training center. Instituted a co-op program within the first year enabling students to gain invaluable field experience. This program, the first of its kind, was adapted to other curricula and is still in use today.
- Received numerous letters of recognition from administration for efforts in coaching students for vocational competitions. First teacher to have a first place winner statewide - - this same student went on to place third nationwide. Subsequently, another student was #1 in PA.

EDUCATION

UNIVERSITY OF PITTSBURGH SCHOOL OF DENTAL MEDICINE, PITTSBURGH, PA
Masters' Equivalent in Vocational Education. Certificate in Dental Assisting - Oral Hygiene Curriculum

CALIFORNIA UNIVERSITY OF PENNSYLVANIA, CALIFORNIA, PA
B.S. in Education

CONTINUING EDUCATION … Microbial Irrigation and Instrument Sharpening Seminar

25—ELECTRICIAN

JOHN P. HEAPHY

15 North Mill Street
Nyack, New York 10960
(914) 555-3160 • Home
(914) 555-3161 • Fax

Electrician/Project Manager

SUMMARY OF QUALIFICATIONS

An accomplished electrician with extensive project management experience. Background encompasses demonstrated knowledge of residential and commercial site and structure development. Excellent communication and motivation skills with clients, crew and sub-contractors, providing leadership by example. An exceptional work ethic, committed to guiding all projects through to successful completion.

AREAS OF STRENGTH
• ESTIMATING • CONTRACT NEGOTIATION
• PROJECT MANAGEMENT • BID SOLICITATION/ANALYSIS • SCHEDULING

PROFESSIONAL WORK EXPERIENCE

RAPID ELECTRIC - Tarrytown, New York • 1986-1994 & 1996-Present
Project Manager (1996-Present)
Manage the development residential electrical accounts with general contractors for custom homes in Westchester County, New York. Directed electrical phases of rebuilding project after the damage due to Hurricane Marlin. Handled insurance negotiations, estimating and contract negotiations with subcontractor and suppliers. Coordinated overseas shipping of materials and provided on site project management.

Selected Accomplishment:
Directed two separate projects on St. Croix in the U.S. Virgin Islands after the island was devastated by Hurricanes in 1989 and 1996. Completed all assignments on time and within budget.

Electrician/Foreman (1986-1988)
Coordinated job progress for electrical crew responsible for all installation and repair wiring.

UNICORN CONTRACTING CORPORATION - Millwood, New York • 1994-1995
Project Superintendent
Supervised all phases of electrical installation in luxury high-end homes. Supervised all site work on projects; scheduled projects; negotiated contracts; handled change orders and job work orders. Coordinated all aspects of projects with general contractors, engineers, architects, municipal officials and code officers.

Selected Accomplishment:
Reduced overall project costs by 27% by negotiating supply costs.
Generated $250,000 in new business.

EDUCATION
• STATE UNIVERSITY OF NEW YORK AT DELHI
Associate of Science • Electrical Technology, 1986

LICENSES
• **Licensed Electrical Contractor**
State of New York and the U.S. Virgin Islands

JEFFREY M. (MIKE) JOHNSON

925 SUNNYDALE DRIVE, CHICAGO, IL   123.555.7890

ENTRY LEVEL MANAGER - ENVIRONMENTAL OR RECREATION FIELD

B.A. Park/Recreation Management / Hospitality, Activity Coordination, Public Relations Experience
6 Years Transferable Service And Quality Assurance Experience - Retail and Manufacturing

DEMONSTRATED KNOWLEDGE

• Community Recreation And Revitalization Programs • Environmental Studies • Grant And Proposal Writing •

HIGHLIGHTS AND EXPERIENCE

Highly effective title. Entry-level position being sought is clearly defined and is uniquely joined by education and a synopsis of work experience.

Highlights and Experience is an effective way to showcase talents and achievements for recent college graduate with a modest work history.

- Emerged as #1 activities intern while with a world-renowned ski resort. Hosted scenic tours and provided tourist information at booths and on toll-free lines. Coordinated 3 special events. Planned and implemented corporate events and children's activities. Published daily activity newsletter. Gained insight into resort operations and the importance of rigorous service standards.

- Assisted in developing a master plan for a community park. Project addressed the need to diversify activities to provide additional incentives for its 5 member communities while creating additional revenues. Suggested 5 low-cost fundraising activities increasing public awareness which, when implemented, raised $250,000.

- Co-authored a grant proposal for harmonious use of recreational acreage and existing wetlands at a local park. Funding addressed wetlands and proposed upgrades to walking and bicycle paths, parking lots and landscape, with added youth activities. Suggestions were integrated into final proposal awaiting approval.

- Spearheaded wetlands delineation study, gained understanding of legal and environmental considerations, as well as analytical processes involved in assessing and maintaining wetlands.

- Volunteered for a day-long project with the US Army Corps of Engineers. Used recycled Christmas trees as part of a stream rehabilitation creating hatcheries for fish and aquatic wildlife.

- Core coursework includes Developing a Master Plan, Developing Managed Leisure Activities, Hospitality Industries, Program Planning Administration, Recreation Industry Management, and Site Planning Design.

- Nearly 2 years in sales, merchandising and inventory management for a major retail chain.

- 4 years of transferable knowledge of production, molding, finishing, assembly, QA/QC inspection and documentation while assisting a multi-million dollar manufacturing concern.

- Computer literate, extensive work on IBM compatibles. Windows 95, WordPerfect and Word. Direct experience with e-mail, the Internet and World Wide Web queries.

CUSTOMER SERVICE/STOCK PERSON ... PETS GALORE, ANYWHERE, USA   **1996 to Present**
Direct Customer Contact - Diverse Clientele - Inventory Coordination - Busy Retail Location

ACTIVITIES INTERN ... SOME MOUNTAIN RESORT, MOUNTAIN HILLS, PA   **Summer, 1997**
Public Relations - Special Events and Activities Coordination - High Visibility Position

ASSISTANT ... A UNIVERSITY IN PENNSYLVANIA, COLLEGETOWN, PA   **1996 to 1997**
Foreign Language Lab - Administrative Assignments - High Profile Position

ASSEMBLER / MOLDER ... INSTRUMENTATION INC., NEAR COLLEGETOWN, PA   **1992 to 1996**
Medical Device Manufacturer - Assembly - Quality Control Inspection and Documentation

EDUCATION

A UNIVERSITY IN PENNSYLVANIA, COLLEGETOWN, PA   **1997**
Bachelor of Arts in Park and Recreation Management

ACADEMIC ACTIVITIES AND HONORS

Dean's List for Academic Achievement. Student Activities Board (SAB). University Student Ambassadors.
Rotaract Club. US Army Corps of Engineers Volunteer.

27—ENVIRONMENTAL ENGINEER

Ellen Stafford

An extensive, highly effective two-page resume with powerful introduction that includes Qualifications, Regulatory Affairs and Project Management Experience.

Resume is results-oriented backed by bulleted achievements and contribution in the Experience section.

18 Shadow Crossing
O'Fallon, Illinois 62269
(618) 555-3333

Environmental Engineer/Project Manager

Seven years' experience managing all phases of environmental projects including hazardous waste management, pollution prevention, natural/cultural resources preservation, soil treatment technologies, and air/water quality. Directed numerous remediation investigations and cleanup actions, reducing exposure and liability, cutting costs, and achieving regulatory compliance.

Qualifications include:

Phase I/Phase II - Site Assessment/Remediation
Crisis Management & Emergency Response
Government Liaison/Public Affairs

Wetlands Protection/Mitigation
Project Budgeting and Management
Contract Negotiations

Regulatory Affairs

Extensive knowledge of Superfund requirements and government regulations:
RCRA, CERCLA, SARA, NCP, UST, CWA, OSHA, and NEPA

Project Management Experience

Directed project teams of up to 20 engineers and field personnel. Wrote project investigation, remediation and management plans, prepared budgets, and monitored field operations. Oversaw industrial hygiene standards, occupational safety, health and safety affairs, and permitting. Managed outside liaison affairs with contractors, regulatory agency personnel, and media. Coordinated base closure and property conversion/disposal actions.

Professional Experience

Environmental Engineer/Project Manager, 1994-1997
McGuire Air Force Base, New Jersey

Scoped, initiated, and negotiated remedial investigation/feasibility study project at three former underground storage tank sites. Served as Quality Assurance Evaluator during field work and work plan preparations.

- As Natural and Cultural Resource Manager, finalized basewide wetlands delineation, threatened and endangered species survey, and historical and archaeological inventories.
- Drafted all wetlands permits and associated Pinelands development applications per state law.
- Initiated rapid response (emergency) project resulting in notice of violation avoidance and savings in excess of $850,000. Successfully coordinated work plans with Dept. of Environmental Protection and Pinelands Commission.

Environmental Engineer/Installation Restoration Program Project Manager, 1994
March Air Force Base, California

Identified, analyzed, and formally established IRP objectives to achieve remedial actions/cleanups. Developed estimates, justifications, and facilitated rapid review and subsequent acceptance of planning/design documents by federal, state, and local regulatory parties. Reviewed contractor work for environmental/health/permit compliance.

- Monitored and oversaw extensive contaminated soil and landfill removal actions at four sites.
- Directed contractors during emergency response removal actions in sensitive wetlands and endangered species habitat.
- Coordinated removal actions of underground storage tank removal, landfill excavations and capping, and drum burial site removal.

Professional Experience (cont.)

Environmental Engineer/Installation Restoration Program Project Manager, 1993-1994
Air Force Base Conversion Agency, Norton Air Force Base, California

Provided on-site representation prior to and following base closure. Prepared, implemented, and maintained IRP and Environmental Compliance and programs to abate and limit contamination of the environment and endangerment to public health. Served as spokesperson to public and private entities on the administrative and regulatory requirements under CERCLA, SARA, and RCRA.

- Led all post-closure environmental compliance programs including storm water permits, hazardous waste disposal, RCRA facility closures, and underground storage tank removal.
- Finalized the TCE Groundwater Plume Record of Decision with extensive regulatory input/coordination.
- Accomplished an EPA requested basewide Preliminary Assessment/Site Investigation at 43 areas of concern.
- Authored NEPA documents supporting the lease and transfer of approximately 10 land parcels.
- Provided engineering review on contracts in excess of $4 million.
- Streamlined IRP comprehensive budgeting and billing process.
- Assisted in conversion of the Technical Review Committee to the Restoration Advisory Board.

Environmental Engineer/Remedial Program Manager, 1990-1993
Minot Air Force Base, North Dakota

Represented Minot AFB as the manager of a $5 million restoration program. Implemented remedial investigation at petroleum, oil, and lubricant facility. Completed two landfill site investigations.

- Conducted remedial action quality assurance evaluation of basewide underground storage tank removal, soil cleanup, and RCRA landfill closure.
- Successfully gained regulatory acceptance and closed eight of eleven IRP sites in FY91, the highest total in Strategic Air Command.
- Created and led the original Technical Review Committee and conducted quarterly meetings.

Additional background:

Associate Engineer, City of Merced, California, 1990: Created preliminary design for replacement and expansion of existing sewer system. Completed survey and design on storm drainage, ditches, culverts, and piping.

Transportation Engineer, Georgia Department of Transportation, 1988-1990: Completed department-wide two year training program. Inspected multi-million dollar highway construction. Provided baseline data for environmental impact analyses. Designed a compatible computer program to modernize existing traffic planning program.

Education/Specialized Training

Bachelor of Civil Engineering, Georgia Institute of Technology, Atlanta, Georgia, 1988

- Hazardous Waste Operations and Emergency Response Training (Refresher), 1994/1995/1996
- New Jersey Freshwater Wetlands Permitting, 1995
- Technical Requirements - Site Remediation, 1995
- USAF/EPA Team Approach to Environmental Cleanup and Risk Communication, 1992
- Hazardous Waste Operations and Emergency Response Training, 1992
- Treatment Technologies for Superfund, 1992
- Installation Restoration Program, 1992

Knowledge of Windows '95, Word for Windows, PowerPoint, Excel, WordPerfect, Wang Systems, Computer Aided Design-Intergraph System, GIS Systems, ENVEST, RACER, and ADPM.

28—FLORIST/STUDENT

Janice W. Reynolds
ENTRY-LEVEL FLORIST / FLORAL DESIGN / LANDSCAPE ARCHITECT

Introducing a highly enthusiastic and creative individual seeking to successfully blend academic training in floral / landscape design with more than 10 years practical experience supported by solid professional references.

Education

TEXAS TECH UNIVERSITY, Lubbock, TX
Bachelor of Science: Landscape Design and Management (GPA 3.6), 1998

AUSTIN COMMUNITY COLLEGE, Austin, TX, 1995
Associates in Science: Landscape Design (GPA 3.56), 1996

* Member: Texas Tech Gardening Club and Texas Tech Environmental Association
* 2nd Place Finish: Annual Floral Pageant, 1997 and 1998 (out of 170 participants)

Internship

GOLDEN GARDENS, Neally, TX   1997 & 1998
Floral/Landscape Intern
A two year program that included floral design, interior and exterior landscape design, equipment utilization including tractor, tiller, and related small equipment, planting, product merchandising and display, tree and exotic plant care/ maintenance, and customer service and support.

Employment (While Attending School)

REGENTS FLORISTS, Lubbock, TX   1997 & 1998
Assistant Florist
Worked with owner and customers in designing and preparing floral arrangements for weddings and other large scale events. Customized each arrangement to customer's specific needs and budgets. Assisted in ordering products and related non-floral products including vases and baskets.

GOLDEN LEAF FLORISTS, Houston, TX   1988 - 1994
Assistant Florist
Worked in family-owned business since the age of 14 in all areas of floral design, purchasing, inventory control, facilities management, customer service and support.

*1801 Westshore Drive * Irving, TX 75061 * (970) 555-9102 * e-mail: gardenbum@aol*

Sierra Beane

15 North Mill Street • Nyack, New York 10960 • (914) 555-3160

OBJECTIVE

Seeking a challenging and growth oriented position in the food service industry utilizing extensive educational background.

> *Recent graduate effectively highlights education and credentials atop the resume.*
>
> *Areas of Strength quickly portray her five areas of skills.*
>
> *Affiliations appear in the middle of the resume indicating commitment to "new" profession.*

EDUCATION

B.O.C.E.S • Nyack, New York
Culinary Arts • GPA 4.0 • June 1998
Curriculum Highlights:
Culinary Arts • Restaurant Management • Sanitation • Nutrition • Menu Design • Purchasing • Hospitality Supervision • Bar & Beverage Management • Management by Menu • Controlling Costs • Food Service

AREAS OF STRENGTH

Plate Presentation • Sautéing • Dessert Plating • Selected Regional Cuisine • Sauces

PROFESSIONAL AFFILIATIONS

American Culinary Federation • *Member*

EXPERIENCE

Rigatoni's - Tallman, New York • 1997/Present
Sous Chef
Prepare stocks, sauces and sautés for this northern Italian restaurant serving 125 covers daily.
Selected Accomplishments:
*   *Promoted from line cook to Sous Chef with two weeks due to superior levels of performance.*

Pearl River Hilton - Pearl River, New York • Spring/Summer 1997
Student Extern

Prepared cold appetizers and salads, dessert plating and set up stations. Handled sautéing, braising and poaching for the catering division of this upscale hotel serving up to 400 guests at special events.

30—FOOD AND BEVERAGE MANAGER

Comprehensive one-page resume with lots of information.

First half of the resume communicates diverse areas of strength supported by the Professional Experience section.

Resume is achievement-oriented as indicated by five key "RESULTS."

MARCUS H. FRANCESCA

5437 East 42nd Street • New York, New York 10012 • Phone (212) 555-5421

MULTI-SITE FOOD & BEVERAGE OPERATIONS MANAGER

Dynamic 10+ year professional career leading daily operations of fast-paced, F&B operations within the entertainment, hotel, restaurant and contract food service industries. Strong qualifications in personnel development, team building and team leadership. Effective motivator, trainer and mentor. Dedicated to continuous improvements in quality, productivity, efficiency and customer service. Core competencies:

Operations Management
Scope of responsibility is diverse and includes planning, budgeting, expense control, recruitment, staffing, scheduling, procurement, inventory control, menu planning/pricing, facilities management, and customer service/guest relations. Direct and decisive with "hands-on" management style. **RESULT: Contributed to solid cost reductions and revenue/profit growth.**

Customer Relations
Manage ongoing relationships with key customer accounts. As on-site liaison to client management, coordinate the planning, development and delivery of customized service programs. Advise regarding customer contract negotiations and the allocation of personnel, budgets and resources. **RESULT: Recognized throughout career for outstanding customer service and relationship management skills.**

Human Resources & Training
Direct staffs of up to 80 responsible for food preparation and service delivery. Plan staff schedules to ensure adequate manpower coverage, coordinate employee training, and design/implement incentives and other motivational programs to enhance customer service competencies. **RESULT: Consistently improved and strengthened customer relations/retention.**

Purchasing, Vendor Relations & Inventory Management
Plan, budget and manage all purchasing, inventory planning and stock replenishment programs. Concurrently, source and select vendors, negotiate terms and conditions, and implement vendor quality standards. **RESULT: Consistently maintained costs at or under budgeted projections.**

Budgeting & Financial Affairs
Participate in the planning, development, administration and management of annual operating budgets valued at up to $300,000 annually. Evaluate personnel, supply, equipment and material requirements to assist in budget planning and forecasting for multiple operating locations. **RESULT: Consistently managed operations to within 97% of budget.**

PROFESSIONAL EXPERIENCE

Manager – AT&T World Headquarters, International Food Management Services (1996 to 1998)
Recruited to manage $750,000 annual contract providing cafeteria and vending services for 1500 employees at AT&T World Headquarters. Challenged to improve menu selection, introduce professional service delivery and provide operations expertise to enhance client relations where previous management had failed.

Banquet Captain / Maitre'D, Waldorf Astoria Hotel (1990 to 1995)
Directed special event and banquet affairs for this prestigious hotel. Managed an average of 60 events per month including the "Mayor's Breakfast," a benefit for 200 regional business leaders who played a key role in bringing new business opportunities to metro New York.

Assistant Manager, Sands Hotel & Casino (1983 to 1990)
Fast-track promotion through increasingly responsible operating management positions from Cook to Supervisor to Assistant Manager. Directed the operations of up to five locations with supervisory responsibility for 80+ employees serving more than 2500 clients daily.

EDUCATION

B.S. in Fine Arts & Business Management, Culinary Institute of America (1983)
Apprentice Chef under Chef Garrett Winchester

31—FORKLIFT OPERATOR

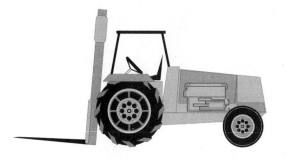

Graphic will definitely separate this resume from his competitors.
Good use of listing Areas of Strength up front.

WAYNE C. COOLIDGE
18 Builtmore Lane
Las Vegas, NV 70623
(505) 555-1234

FORKLIFT OPERATOR
WAREHOUSE OPERATIONS / SHIPPING & RECEIVING

More than 4 years experience as a forklift operator, experienced in all areas of warehouse operations, shipping and receiving. An impeccable safety record, recognized for consistently doing "more than what's expected."

AREAS OF STRENGTH

Forklift operations
OSHA safety regulations
Inventory control
Training and supervision
Quality control

Shipping & receiving
Traffic management
Cycle counting
Documentation
Customer service and support

EMPLOYMENT

<u>Las Vegas Beverage and Distribution Center</u>, Las Vegas, NV   1993 - Present
Warehouse Supervisor / Forklift Driver

- Began working in warehouse as a stocking clerk while completing high school in 1993.
- Promoted to stocking supervisor, forklift driver, and warehouse supervisor as a result of outstanding work.
- Assisted in renovating 50,000 square foot distribution center between 1994 and 1996.
- Train and supervise 7 full and part time employees.
- Operate Huchinson deluxe UA2000 Forklift - trained by prestigious Huchinson Training Institute.

EDUCATION

Las Vegas Vocational High School, Las Vegas, NV
Graduate, 1996

* References and Safety Record Furnished Upon Request *

32—FUND-RAISER

RACHAEL PETERS
1776 Spirit Way
Seattle, WA 91000
(606) 555-2468

Skills:

* Raising money
  * Budgeting & cost control
    * Raising money
      * Employee management & supervision
      * Raising money
    * Strategic program planning
  * Raising money
* Maximizing donations for direct giving

Achievements:

* Raised over $4.2 million in past 5 years for the United Way & Save The Children.
* 5-year consistent record of attaining projected fund raising targets.
* Reduced volunteer turnover of 38% to less than 8% annually. Increased donations 123%.
* Play pivotal role in assuring highest % of donations going to program - 81% (STC).

PROFESSIONAL FUND RAISING - PROGRAM DEVELOPMENT

NATIONAL / INTERNATIONAL CHARITIES & FOUNDATIONS

Employment:

United Way, Seattle, WA
Regional Coordinator   1981 - Present
Fund Raiser

Save The Children, Seattle, WA
Fund Raiser   1988 - Present

Education:

Wharton School of Business - U. of Penn.
MBA with Honors, 1979
B.A. in Organizational Behavior, 1975

Affiliations:
Member: United Way, Save The Children, Special Olympics, March of Dimes

Management Style:
* Team Player - Goal Oriented
* Seasoned Trainer & Motivator
* Highly Ethical

Personal & Professional References
& Supporting Documentation:
Furnished Upon Request

33—FUND-RAISER

HOLLY S. LEEMAN
800 Bates Road
West Palm Beach, Florida 33405
(561) 555-5604

FUND-RAISING PROFESSIONAL

A dynamic top-performing **Development Professional** experienced working with high net-worth individuals, corporations, and the general public to secure major/planned gifts since 1990. Successful in donor prospecting, cultivation, and recognition.

AREAS OF STRENGTH

Organizational/time management skills	Presentation/public speaking skills
Leadership skills	Networking/partnership building
Media relations	Strategic planning
Volunteer training and development	Budgeting management
Key donor development/management	Writing/detailed report generation

PROFESSIONAL EXPERIENCE

HOSPICE OF PALM BEACH COUNTY, INC., West Palm Beach, Florida   1994 - Present
Director of Development
Full responsibility for all aspects of fund raising including planned/major gifts, direct mail, special events, and grant writing management. Chair Research and Development Project. Appointed to and serve on Marketing, Communications, and Outreach Committees. Member of Senior Leadership Council. Named Alternate Spokesperson for the agency.

- Increased annual fund raising efforts from $1.4 million to $2.1 million - a total of $6.8 million in a 48-month period.
- Conceptualized/implemented innovative and successful direct mail marketing program - raised $400,000 in 4 years.
- Managed/directed three auxiliary revenue-generating guilds resulting in an additional $500,000 in revenue.
- Restructured Development Department increasing efficiency in donor relations, recognition, and. acquisition.
- Spearheaded and administered FundMaster software program - trained all staff members (20+ people).
- Grew donor base file from 20,000 to 68,000 between 1994 and 1998.

ECCLESTONE ORGANIZATION, West Palm Beach, Florida   1993 - 1994
Director of Community Relations
Authored comprehensive community relations plan to position the company for enhanced public exposure and to maximize business development and growth opportunities.

- "Executive on Loan" to the Economic Council of Palm Beach County for Operation Headquarters (9/93-1/94).
- Developed procedural guidelines for Operation Headquarters project - including special events and publicity.
- Identified national CEO prospects interested in Corporate Headquarters' re-location to Palm Beach County.

REPUBLICAN PARTY OF PALM BEACH COUNTY, West Palm Beach, Florida   1990 - 1993
Executive Director
Executed political visitation/press conferences for members of the U.S. Senate and House of Representatives, Florida Congressional Delegation, Federal/State Cabinet members, and other ranking political leaders. Liaison to the Secret Service, FBI, White House, State and House Legislatures, Constitutional Officers, County Commissioners, and Gubernatorial Candidates.

- Managed daily operations of 100,000 member organization and created/directed annual member drive.
- Developed student intern program and training program for volunteers.
- Received Special Recognition Award for outstanding contributions - 1992 and 1993.
- Spokesperson for organization. Co-authored county political plan.

33—FUND-RAISER (*CONT.*)

PROFESSIONAL EXPERIENCE (continued)

SELIG CHEMICAL MANUFACTURING CO., Atlanta, Georgia   1987 - 1990
Sales Representative / Supervisor - Palm Beach County Territory
Responsible for developing Palm Beach County territory for major subsidiary of Fortune 500 Company. Assisted in the development of national training seminars for new sales professionals. Featured speaker at national and regional conferences and seminars. Offered Regional Director position to travel the Southeastern United States to recruit and train new sales personnel.

- Ignited sales from $0 to $750,000 in less than 24 months.
- #1 Salesperson, 1987-88; most new accounts, 1988; top order producer, 1989.
- Member 110% sales club, 1989, 1990, Simon Selig Award, 1990.
- Broke all existing sales records in 1987 and 1988.

CHENEY BROTHERS, INC., West Palm Beach, Florida   1983 - 1988
Customer Service Manager / Executive Assistant to the President
Promoted from Customer Service Representative to Executive Assistant within three months and to Customer Service Manager in less than 24 months for food service distributor with annual sales exceeding $25 million.

- Serviced key accounts including the Breakers Palm Beach, Boca Raton Resort and Club, and the Hyatt Hotel.
- Instrumental in facilitating/closing add-on sales for commercial clientele.

EDUCATION & SELECTED TRAINING

BBA Degree Northwood University, West Palm Beach, Florida (Expected date of graduation, 5/99)

SEMINARS & WORKSHOPS
- Major Gift Planning - Robert F. Sharpe
- Planned Giving - Robert F. Sharpe
- Institute of Planned Giving at the College of William and Mary - Robert F. Sharpe
- Taking the Fear Out of Asking for Major Gifts - Jim Donovan
- How to Ask For Money; Making Strategic Choices at Critical Development Stages - Nancy L. Brown
- How to Make Presentations With Confidence & Power; Project Management; Communication Skills - Fred Pryor
- Presenters School - Jenna Commander
- Facilitator Training - Ralph Parilla
- How to Handle People With Tact & Skill; CareerTrack
- Coaching and Team-building Skills for Managers and Supervisors - Skillpath
- The One Minute Sales Person - Larry Wilson

- COMPUTER SKILLS: Intermediate and advanced Word and WordPerfect; Quattro Pro; FundMaster; Internet

CURRENT COMMUNITY INVOLVEMENT/PROFESSIONAL AFFILIATIONS

National Society of Fund Raising Executives - Member (1994-Present), Board Member (1995-97); President Elect (1997)
Planned Giving Council of Palm Beach County - Member (1995-Present)
Palm Beach Chamber of Commerce - Member (1994-Present)
Women in Communications - Member (1996-Present)
Palm Beach County Community Relations Group - Member (1995-Present)
Florida Hospices, Inc. - Co-chair (1995); Development and Marketing Conference (1995-96)
El Cid Neighborhood Historical Association - Member (1995-Present)
Governor's Club - Member (1998-Present), Board of Ambassadors (1998-Present)
Executive Women of the Palm Beaches - Member (1992-Present), Board Member (1995-97); Co-chair Leadership Development Committee (1996-97)

■ *References Furnished Immediately Upon Request -*

34—GENERAL MANAGER

Coleen S. Kubeck
21 East 39 Street, Apt. #3
New York, NY 10016

Senior-level General Management ■

Executive Profile

■ Over ten years of senior-level experience in astute business analysis and profitable management of 20 million dollar custom manufacturer. Forecast sales trends, enhance revenue streams, turn around troubled operations, and achieve profitability in down-trending markets. Administer all manufacturing, marketing, and environmental control functions. Supervise staff of up to 25 direct and indirect reports.

■ A hands-on manager and critical thinker who can learn quickly, develop expertise, and produce immediate contributions in systems, analysis, business operations, and motivational team-management. Possess a valuable blending of leadership, creative, and analytical abilities that combine efficiency with imagination to produce bottom-line results.

Proven Areas of Knowledge

■ business planning / development	■ operations management	■ operational troubleshooting
■ revenues and margins	■ multiple project management	■ task analysis
■ modular manufacturing architecture	■ facilities management	■ capital / consumable purchasing
■ trend and competitive analysis	■ crisis management	■ high-expectation client relations
■ joint venture formation	■ environmental management	■ training and development

Executive Highlights

■ Produced exceptional company growth, increased gross margin, enhanced productivity, and set new quality standards through the proactive design of innovative programs, sales techniques, and manufacturing methodologies as well as the imaginative use of unique suppliers.

■ Doubled company's accounts and quadrupled sales by design and implementation of profitable value-added services. Division generated unit sales per employee that were 1.5 times that of the industry's largest independent service bureau. Produced division turnaround time of 10 to 12 weeks vs. industry average of 13 to 18 weeks.

■ Created an innovative in-house service bureau, the only one of its kind in the industry. Produced new revenue streams representing 12.5% of sales through pro-active marketing of new service bureau as a quasi-independent operation that gained new cross-industry, non-printing accounts.

■ Avoided purchase of multi-million dollar computer graphics system through inventive utilization of offset and reprographic service bureaus, reducing expenses even further by scheduling projects in bureaus' down-times.

■ Designed and directed company's strategy to prevent major losses from 40% erosion of customer base during late 80's recession. Spearheaded production of proprietary nationally distributed wallcovering collections, advertising, and collateral materials. Developed manufacturing relationships with large furniture producers and cosmetics firms. Marketed color separation and film services to competitors.

■ Prevented massive disruption in service by assuming immediate control of all manufacturing operations in crisis response to key administrators' mismanagement and subsequent departure.

Coleen S. Kubeck

Employment History

Distinction Printing, Ltd., Long Island City, NY   1980 to present

Company manufactures custom wallcoverings, decorative laminates, large-scale graphics and point-of-purchase specialties, with a peak sales volume of $20 million and a number two ranking in this specialized industry of twenty contract wallcovering printers in the United States.

Vice President and General Manager   1995 to present

Assumed control of all manufacturing operations following departure of two key managers. Situation required immediate action to position company to recover from mismanagement. Downsized staff, slashed overhead, cross-trained personnel, instituted strict housekeeping controls to curtail waste, and reduced inventory — all with no reduction in quality.

Increased margins by 60% on four existing customer collections, signed with two new national distributors, and developed high margin accounts. Increased business without increasing expenses through the use of vendors' sales people as a de-facto sales force to market company's services to non-competing screen-print industries.

Manage facility operation and safety / environmental coordination. Supervise all facility functions including hiring of contractors for maintenance and renovations. Administer vital hazardous materials program for entire corporation, including training, compliance, documentation, and reporting. Directed all repairs, contractors, and insurance affairs after partial roof collapse and flooding in 1996.

Sales and Operations Manager   1987 to 1995

Held full P&L responsibility. Increased division sales by 30% between 1987 and 1990 and steered company through the recession of the late 80's / early 90's when fully 40% of the industry's customer base was lost through consolidation and bankruptcy. Researched and developed new markets, created new opportunities as outsourced producer for competitors' small runs, and positioned company as fast-turnaround specialist.

Pre-Press Manager   1980 to 1987

Planned production, scheduled, procured consumables, capital equipment, and outsourced services. Created value added services that directly contributed to doubling of company's account base and quadrupling of sales from 1980 to 1987. Ran division turnaround times typically 30% less than industry standard, with no decline in quality.

Education and Development

Masters of Business Administration, New York University, New York, NY, 1990

Bachelor of Arts in Business Administration, State University of New York at Stony Brook, 1979

Technology

Use PC word processing, data base and spreadsheet software (MS Office), the Internet and E-mail.

Easily learn specific industry systems and software. Familiar with Mac, especially graphic arts software.

Everett Morris

50 Trouble Drive
Fairview Heights, Illinois 62208-2332
(618) 555-8228

Graphic/Visual Arts Specialist

Seventeen years experience managing and coordinating visual arts projects from concept to completion. Accustomed to working on multiple projects with short notice, little or no instruction, and total creative judgment for quality and details of finished product.

- Well-rounded business management skills, with proven ability to match customer needs with a wide variety of graphics, visual arts tools, and approaches.

- Effective at communicating ideas and capturing the interest of the intended audience.

Professional Experience

Visual Information Specialist   1989-Present
(self-employment under contract to the Defense Information Technology Contracting Office, a full-service telecommunications and information systems procurement office with 400+ employees supporting the entire DoD and 56 non-DoD organizations)

Managed all operations of a $100,000 business supporting telecommunications, information systems, administrative, financial, and management staff. Responsibilities included record keeping, accounting, budgeting, and inventory control.

- Prioritized customer requirements and assigned workload to meet changing contract specifications and customer deadlines.

- Provided detailed instructions to employees for new, difficult or unusual requirements. Ensured quality of completed products prior to delivery to customers.

- Served as a concept consultant to assist and advise customers on colors, content, and size of finished products.

Additional Experience (1979-1989): Held the same title of Visual Information Specialist as a contract employee without the business management responsibilities.

Education

Graphique Commercial Art School, 1978-1979
St. Louis, Missouri

General Coursework, Belleville Area College, 1976-1978
Belleville, Illinois

35—GRAPHIC ARTIST (*CONT.*)

Creative and Technical Skills

Paste-Up

Designed and produced paste-ups and artwork for illustrative slides, viewgraphs, and charts used in formal presentations, static displays, cover brochures and posters used for briefings to senior military and civilian personnel, static displays at trade shows, and various reports.

Engraving/Signmaking

Designed and produced plastic and metal engraved signs and plates used for organizational awards, plaques, and other recognitions. Designed and produced vinyl/plastic signs for badges, nameplates, name tags, cubical identification, and hallway directories.

Audiovisual Displays

Maintained audiovisual equipment library including the distribution, setup and operation of video and still cameras, audio recorders, overhead transparency, 16mm and 35mm projectors for presentations and training sessions.

Computer/Graphic Design

Designed and produced slides, charts, and graphs for weekly staff meetings and formal presentations using various software packages such as Harvard Graphics, Freelance Graphics, and PowerPoint. Manipulated color and size of clip art with software packages such as Arts and Letters. Produced forms, flyers and publications using PageMaker and Corel Draw software.

Presentation Planning

Provided expert consulting service to assist customers in presentation planning for the various types of media used to market, publicize and document the organization's goal and services. Used Harvard Graphics and Microsoft PowerPoint extensively.

Photography/Video

Provided video/audio and still photographic coverage of formal ceremonies, training, presentations, and satellite downlink training broadcasts. Produced working copies for approval by the customer and master and distribution copies of all materials. Provided photographic darkroom development of film, photo enhancement of prints in both black and white and color.

JENNIFER R. DOUGLAS

1808 Duncan Way
Nashville, TN 37211

AWARD-WINNING HAIR STYLIST
Licensed Cosmetologist

A dynamic, highly creative, and seasoned hair stylist professional offering 16 years of award-winning hair design in major markets including Los Angeles and Boston. Recognized for successfully blending outstanding hair styling techniques with strong business development, customer service, and client retention management skills.

HIGHLIGHTS OF EXPERIENCE

♦ Recipient of the REGAL Award for outstanding styling and hair design, 1992 - 1998
♦ First place in regional styling competition sponsored by Paul Mitchel products, 1997 and 1998
♦ Instructor/consultant for more than 80 individual hair stylists and 14 hair salon owners between 1987 and 1998
♦ A verifiable track record for solid networking and marketing skills in building a strong client base in multiple markets
♦ Consistently generate client bookings exceeding $2,100 weekly in both the Los Angeles and Boston markets
♦ Provide back up management and sales/marketing/promotional support to beauty salon owners
♦ An excellent trainer and mentor to new, up-and-coming hair styling professionals

PROFESSIONAL EXPERIENCE

JEAN PIERRE RAFFAEL'S HAIR CLINIQUE, Beverly Hills, CA	1992 - 1998
SeniorHair Stylist / Public Relation & Marketing	
PAMPER YOURSELF, INC., Beacon Hill/Boston, MA	1987 - 1992
Senior Hair Stylist/ Make-Up Artist	
DIAMOND BEAUTY SALON, Arlington, MA	1985 & 1986
Internship: Hair Stylist/ Make-Up Artist	

EDUCATION & TRAINING

MASSACHUSETTS BARBER AND HAIR STYLING COLLEGE, Brookline, MA
Master Hair Styling License, 1986

Courses and Additional Training :
- Advanced Hair Styling Techniques, Los Angles Hair Styling School, Los Angeles, CA, 1989
- Certificate of Completion: Cosmetology, Northeast Technical School, Boston, MA, 1990
- Hair Replacement Week-long Seminar, Danielle Wilcox School for Hair Replacement, Burlington, VT, 1995
- Hair Replacement Refresher and Update Seminar, Danielle Wilcox School for Hair Replacement, Burlington, VT, 1997
- Advanced Hair Coloring Techniques, Redken Symposium, Las Vegas, NV, 1998

LICENSES

Master Hair Styling and Barber License, California (#103729-AL)
Master Hair Styling and Barber License, Massachusetts (#MS99937586)
Master Hair Styling and Barber License, Tennessee (#67732-BTL)
Licensed Cosmetologist, Massachusetts and California (inactive)
Licensed Cosmetologist, Tennessee (active)

Professional References and Portfolio Furnished Upon Request

37—HEALTHCARE ADMINISTRATOR

JOSEPH E. TEABOL
12015 Shark Infested Drive
Fallbrook, CA92028

HEALTHCARE ADMINISTRATOR / CEO / COO
New Business Start Up / Turn Around Management / Driving Business Growth
Sales & Marketing / Global Vision / P&L Management

- **D**ynamic Operating/Management professional offering 10-plus years of successful leadership positions in healthcare environments. Professional qualifications in strategic planning/implementation, finance management/ budgeting, and personnel leadership/team-building - maximizing individual and group productivity in highly competitive markets. A proven record for creating high-visibility marketing and business development strategies; able to identify and capitalize upon new and emerging market opportunities in igniting revenues and enhancing profit position. Expertise in Medicare administration and protocol.

AREAS OF EXPERTISE

- Organization, administration, time management
- Insuring Federal and State Compliance
- Developing strategic partnerships and alliances
- Customer service and retention management; quality control
- Finance management, budgeting, expense control

- Multi-unit management
- Executive-level negotiator
- Cross cultural business development
- Policy and procedure formulation
- Profit center realignment/reorganization

PROFESSIONAL CREDENTIALS

IMPACT WEST ASSOCIATES, INC. (IWA), Petaluma, CA   1998 - Present
Chief Operations Officer (COO)
- Regional P&L responsibility directing a multi-million operation consisting of 40 outpatient rehabilitation facilities. Recruited 7/98 with key responsibilities to include: Revenue generation, new business development on a national scale, personnel recruitment, training, and supervision, and developing/executing strategic operations and marketing plans.

COMPREHENSIVE OUTPATIENT REHABILITATION FACILITY, West Palm Beach, FL   1995 - 1998
Administrator / Principal
- Planned and developed Comprehensive Outpatient Rehabilitation Facilities (CORF's) nationwide. Ensured total compliance with Federal, State, and Local regulations. Responsible for formulating and implementing extensive policies and procedures for all locations, personnel staffing, training, team-building, and development, and full charge P&L responsibility to include budget preparation/management, banking relations, and short/long term forecasting. Successfully oversaw clinical services, and acquired/organized Managed Care contracts, CARF and JCAHO accreditation and certification - nationwide.

WOODBRIDGE CENTRE West Palm Beach, FL   1985 - 1995
Administrator / Principal
- Directed operations for high-volume psychological services practice with offices in West Palm Beach, Jupiter, Stuart, and Martin County locations as well. Planned and executed comprehensive marketing strategy to ignite growth and profits. Spearheaded the development and implementation of operational management structures to meet the growing needs of the sales and marketing efforts. Successfully negotiated 48 Managed Care contracts. Recruited, trained and managed staff. Developed three separate profit generating offices.

Joseph E. Teabol
Page Two

PROFESSIONAL CREDENTIALS (Continued)

CENTER FOR FAMILY SERVICES, INC., West Palm Beach, FL　　　　　　　　　1982 - 1985
Staff Therapist
♦ Provided individual, family, and group therapy to children, adolescents, and adults. Networked closely with insurance companies and assisted in the development of Employee Assistance Program (EAP).

PARENT CHILD CENTER OF THE PALM BEACHES, West Palm Beach, FL　　　　1977 - 1982
Group Work / Intake Supervisor
♦ Responsible for implementing and providing individual, family, and group psychotherapy to targeted population. Performed community liaison services and networking. Directed the Center's group and intake staff and developed clinical programs - tripling the size of the clinical program and increasing the sources of non-profit funding by 40%.

EDUCATION & TRAINING

NOVA SOUTHEASTERN UNIVERSITY, Fort Lauderdale, FL
Masters - Healthcare Administration (Currently Enrolled); Expected Graduation Date: 6/99

BARRY UNIVERSITY, Miami Shores, FL
Masters - Social Work; Administrative Tract, 1984

SPRING HILL COLLEGE, Mobile, AL
Bachelor of Science - Psychology, 1977

* LCSW, 1990

PROFESSIONAL AFFILIATIONS

American College of Healthcare Executives
Academy of Managed Care Providers
Mental Health Organization
National Association of Social Workers
Chamber of Commerce

References Furnished Upon Request

38—HEALTHCARE DIRECTOR

Julie Kohler

15 North Mill Street - Nyack, New York 19060 - (914)555-3160

OVERVIEW

Highly professional clinical director who has demonstrated continuous growth, achievements and impressive leadership in the management of complex activities within the healthcare industry. Solid business insight with the ability to ascertain and analyze needs, forecast goals, streamline operations and envision new program concepts. Excellent communication and interpersonal skills serve as the foundation to effectively network, collaborate, negotiate and maintain positive partnerships with physicians, staff and outside healthcare organizations. Proficient in the management of a diverse range of departments, professionals and programs through a complete understanding of the healthcare arena and integrated networks.

AREAS OF STRENGTH

MANAGED CARE • PROVIDER NETWORKS • SYSTEM INTEGRATIONS • QUALITY
IMPROVEMENT COMMUNITY BASED NETWORKS • COMMUNITY RELATIONS

FUNCTIONAL RESPONSIBILITIES

- Assumes full responsibility for all daily activities, functions, and decisions; sets priorities, organizes procedures and creates plans for work flow and delegation.
- Formulates, administers, and monitors capital/operating budgets; applies right sizing and revenue enhancement to establish cost control and revenue generation precedents.
- Conceives designs, plans and administers programs and special projects after thorough assessment of needs; coordinates departments and staff to accomplish goals.
- Successfully negotiates and executes contracts including managed care agreements, and develops integration and alternate delivery systems.
- Solicits community support and utilization through effective marketing and public relation techniques to stimulate and increase image and visibility.
- Motivates and evaluates executive and professional personnel; maintains open channels of communication to result in optimum delivery of service, a positive working environment and rapid business development at the department level.
- Seeks and secures grants and creative funding opportunities to improve community health status through the initiation of programs to target identified needs.
- Maintains a focus on the development and enhancement of the professionalism of employees; provides continuing education opportunities to promote maximum professional growth.
- Creates employee recognition programs to increase morale and productivity.
- Continually scrutinizes short-term and long-range strategies, goals, and mission achievements; institutes new philosophies and objectives.

Julie Kohler • *Page Two*

PROFESSIONAL EXPERIENCE

Saint Mary's Hospital - Bronx, New York 1989/Present
Clinical Director
- *Increased revenues by 120% through proactive marketing campaigns and the establishment of three new revenue sources including Physical Rehabilitation, Occupational and Speech Therapy Department, Mental Health Counseling and Substance Abuse Counseling centers.*
- *Decreased housekeeping costs by 30% while increasing overall staff productivity by 31%.*

REGIONAL HEALTHCARE SYSTEM - Boston, Massachusetts 1987/1989
Assistant Clinical Director
- *Developed a multi-state healthcare system consisting of 549 beds with gross revenue exceeding $85 million.*
- *Acted as resource in development of $9.5 million bond issuance.*

EDUCATION AND CERTIFICATIONS

Central Michigan University - *Fort Meade, Maryland*
Master of Arts - Personnel Administration

La Sierra University - *La Sierra, California*
Bachelor of Science - Business Administration

PROFESSIONAL AFFILIATIONS

Fellow in the American College of Healthcare Executives

Member of the American Institute of Certified Public Accountants

39—HEALTHCARE/HOSPITAL ADMINISTRATOR

Excellent resume with thorough opening section depicting Relevant Competencies (bullets).
Professional Experience section is comprehensive and complete.
Strong ending with Professional Achievements section.

CAROLYN NOVINGER
23423 Avenue K, Fallsbrae, California 92081 • 555 555-1212

• HEALTHCARE EXECUTIVE • PATIENT SERVICES MANAGER • ADMINISTRATOR •

Competent, confident, and compassionate healthcare executive with considerable management experience refined by a formal education and specific expertise in Federal and State employment laws. Proven record of success in recruiting, interviewing, selecting, and training healthcare professionals. Strong background in initiating and implementing leadership techniques that significantly improve employee morale and effectiveness, increase productivity and decrease employee turn over rates. Excellent communications and problem solving skills have dramatically reduced numbers of incident reports, improved relations within hospital departments, increased personnel productivity, improved service levels, favorably impacted customer service, and markedly increased overall quality assurance.

RELEVANT COMPETENCIES

• Compensation & Pay	• Employee Orientation	• Performance Standards
• Computer Literate	• Employee Relations	• Program Development
• Staff Coordination	• Long Range Planning	• Program Management
• Department Budgets	• Organization/Scheduling	• Recruitment/Interviews
• Employee Counseling	• Policies/Procedures	• Staff Development

PROFESSIONAL EXPERIENCE

BAY CITY MEDICAL CENTER, Bay City, California - 1988 to present

Human Resources Coordinator responsibilities include review and coordination of all Human Resources Department policies, systems, and procedures. Coordinate Human Resource activities with all other medical center departments. Conduct research and analysis of current policies identifying areas for improvement. Develop and recommend policies revisions and maintain procedure manuals. Perform systems and procedures analyses to support internal control programs. Provide support for periodic evaluations, special services, and programs relating to Human Resources functions. Participate in hospital CQI projects and TCMC division meetings representing the Human Resources Department. Gather and analyze data from other departments and outside services to assist in the development of new programs or the improvement of existing ones.

Phlebotomy Supervisor responsible for adequate staffing of all positions, and completing all administrative and human resource activities required for interviewing, hiring, counseling, motivating, disciplining, and terminating members of the department. Maintained, coordinated, and implemented new programs, protocols, and policy changes. Provided support, advice, guidance, and education to the center's staff. Maintained effective employee relations as well as liaison and communications with the public, physicians, and hospital staff. Reviewed and revised department protocols and standards, updating policies and procedural manuals as needed. Developed performance standards and performed yearly merit performance evaluations. Ensured Phlebotomy Department services were provided and contained within budget constraints. Coordinated expanded phlebotomy staffing at an off-site Clinical Services Center.

LMNO HEALTH CARE CLINIC, Milford, Michigan - 1985 to 1988

Patient Services Manager responsible for management and supervision of patient services department. Oversaw all phases of recruiting, selection of medical and technical

(Continued on next page)

CAROLYN NOVINGER

personnel, hiring, training, counseling, and staff development. Facilitated patient care from start-up for this multi-physician practice through on-going clinic development. Developed and maintained policy and procedure manuals. Coordinated medical and patient services with physicians, nurses, and department personnel. Liaised with Hidden Valley Hospital and community. Ensured services were performed within budget guidelines and constraints.

PRIOR PROFESSIONAL EXPERIENCE

Laboratory Technician, Hidden Valley Medical Center, Monterey, California
Clinical Technician, Carlinad Osteopathic Health Care, Monterey, California
Laboratory Technician, CHAMPUS Medical Center, Lakeland, Texas

EDUCATION AND PROFESSIONAL TRAINING

National University, San Diego, California
Bachelor of Arts in Behavioral Science - Emphasis: Human Resources Management
Graduated **Magna Cum Laude**

Carlinad Center, Monterey, California
Interpersonal Skills - 1,475 contact hours

Carnegie Institute, Detroit, Michigan
Lab Technician Certificated Course of Study

·PROFESSIONAL ACHIEVEMENTS

- Developed and implemented JCAHO proficiency testing procedures to enhance customer service and ensure quality trained personnel.
- Served on hospital task force committees improving performance standards, mandating hourly rounds by phlebotomy personnel, and armband compliance for patient identification augmented by a performance standard for hospital personnel.
- Implemented staff cross-training to ensure expanded coverage and eliminate overtime.
- Restructured staffing decreasing employee requirements and lowering salary costs by more than 20 percent.
- Developed the Phlebotomy Orientation Program introducing a three-phase, goal oriented training program; increased overall efficiency and competency of the staff.
- Participated in planning and developing an off-site draw station; as well as an alternate draw station within the hospital; ensured timely, convenient, professional service levels.
- Established a training program for College Paramedic Interns and Bay City Home Health Nurses to learn phlebotomy techniques resulting in phlebotomy proficiency and quality work performance.
- Participated in new clinic start-up, which involved equipment purchases, supplied inventory medical safety precautions, room layout, and setup.

RELATED PROFESSIONAL ACTIVITIES

- Member, Advisory Board, Regional Occupational Program (ROP), Monterey, California
- Commencement Speaker, Community Colleges, Monterey, California

Appropriate personal and professional references are available.

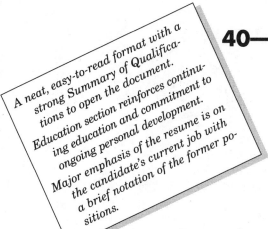

A neat, easy-to-read format with a strong Summary of Qualifications to open the document. Education section reinforces continuing education and commitment to ongoing personal development. Major emphasis of the resume is on the candidate's current job with a brief notation of the former positions.

40—HOME HEALTH AIDE

ALISHA GOFF
Two Cherokee Way
North Reading, MA 01864
(999) 555-1212

SUMMARY OF QUALIFICATIONS

- Well qualified **Home Health Aide** with exceptional clinical nursing skills and strong communication skills complemented by ability to work very well independently and autonomously.
- Excel in professionally managing caseload and setting priorities; strong service delivery and assessment skills enhanced by particularly resourceful style. Key decision-making abilities.
- Develop very strong rapport with patients; highly adaptive, flexible style allows for efficiently and competently working with patient population; exceptional directional and map-reading skills.

PROFESSIONAL EXPERIENCE

VISITING NURSE ASSOCIATION OF BOSTON   1994–Present
Home Health Aide, Evening Shift
- Short- and long-term assignments for heavy patient load comprises many geriatric patients and other conditions typically presenting in an urban environment; extensive direct patient care includes dressing changes, monitoring changes in wound status, and making recommendations regarding types of wound dressings and frequency of dressing changes to maximize wound healing.
- Provide care to diabetic patients; administer insulin, monitor blood sugars, and provide instruction in all aspects of diabetic self-care.
- Other responsibilities include catheter insertion, instruction in self-catheterization, trache care, and disimpactions as well as medication administration and supervision.
- Respond to patients calling in with after-hours condition changes and provide assessments and input to RNs/physicians determining future treatment/intervention.

UNIVERSITY OF MASSACHUSETTS HEALTH CENTER   1992–94
Nurses' Aide, Med-Surg Floor
- Assisted with delivery of direct patient care in post-op environment; orthopaedics concentration.

BOSTON UNIVERSITY MEDICAL CENTER RADIOLOGY DEPARTMENT   1989–92
X-Ray Technician

EDUCATION

FRAMINGHAM COMMUNITY COLLEGE — **Associate in Science Degree, Radiologic Technology** (1989)

Continuing professional education includes ongoing completion of programs in such areas as Trauma Nursing Core Course (1997) and Stabilized Patient Care Training (1997).

CERTIFICATIONS

- Maintain CPR Certification (expir. 26 Oct. 00)
- Emergency Nurses' Association Certification

113

41—HOTEL EXECUTIVE

SUE SENTELL, FMP

SUMMARY

- ■ **Accomplished Hotel Executive** with 20 years' experience in food service field; hold distinguished FMP credential.
- ■ Consistent track record of successfully turning around faltering operations and creating profitability and excellence; utilize keen assessment and problem-solving abilities, dynamic training techniques, and key motivational strategies that build accountability and enhance staff performance.
- ■ Flexible, adaptable style and hands-on approach; a skilled manager who thrives in an atmosphere demanding excellence, autonomy, and strong team-building skills.
- ■ Possess highly polished communication and interpersonal skills.
- ■ Graduate of several of Europe's premier hotel management/culinary institutions; professional management experience acquired through employment with some of Europe's most prestigious establishments as well as a Boston four-diamond hotel; multilingual fluency includes English, German, and French.

PROFESSIONAL EXPERIENCE

1990–Present

Marriott Hotel and Conference Center • Bangor, ME
General Manager
Complete management responsibility for hotel operations with P&L responsibility for Food and Beverage department. Oversee facility including full-service restaurant, room service, lounge, pool bar, grand ballroom, three junior ballrooms, and 14 conference rooms. Overall facility comprises 30,000 sq. ft. of meeting space, banquet facilities servicing up to 1,600 guests, and 214 guest rooms. Manage staff of 200. Hotel revenues exceed $9.5 million, with food/beverage revenues representing $4.7 million per annum.

Select Accomplishments ...
- • Achieved "Hotel Group Food and Beverage Hotel of the Year Award" (1997) on the basis of exemplary performance throughout all facets of operation (guest services, food and beverage, and overall profitability).
- • Named "Marriott Hotels Worldwide President's Award" winner (1995) for significantly exceeding guest satisfaction standards and overall performance measurements for hotel.
- • Overseeing facility-wide renovation ($4 million) scheduled for completion 1999.
- • Implemented comprehensive cross-training program complemented by development of in-depth job descriptions and accountability for all personnel.

Corporate Food & Beverage Consultant (1992–96)
Concurrent with management responsibility for the Marriott, served as Food and Beverage Consultant for hotels managed and operated throughout the United States by the Marriott's holding company, ABS Hotel Associates.

- • Implemented comprehensive cost control systems; facilitated training in menu planning as well as food and beverage marketing.
- • Instructed Tips, ServSafe Sanitation Certification, and Guest Satisfaction Seminars.

1986–90

Boston Hilton Hotel • Boston, MA
Director of Food & Beverage (1988–90; promotion)
Managed entire Food and Beverage department of four-diamond hotel comprising two restaurants (including one of downtown Boston's finest upscale restaurants), lounge, night club, retail bake shop, kosher kitchen, and room service.

114

SUE SENTELL, FMP

Page Two

PROFESSIONAL EXPERIENCE

Boston Hilton Hotel *(cont'd.)*
Hired, trained, and managed staff of 125. Facility included 410 elegant guest rooms as well as 32,000 sq. ft. of meeting space. Annual food/beverage revenues exceeded $5 million.

Beverage Manager (1986–88)

1984–86 **Copenhagen Airport Hotel** • Copenhagen, Denmark
Restaurant Manager
Managed the *Count Dane* restaurant and grillroom.

1981–84 **Hotel des Paris** • Paris, France
Restaurant/Banquet Manager
Managed the *Escoffier* restaurant as well as the banquet department.

1978–81 *Hotel Apprentice* at such fine European hotels as the London Hilton in England, the Cunard Hotel Paris in France, and Holiday Inn Venice in Italy.

EDUCATION

■ **Klesheim Hotel School, University of Salzburg** • Salzburg, Austria
Institute of Tourism and Hotel Management
Diploma in Hotel Management (1981 Graduate) — Scholarship Recipient

■ **Kilburn Polytechnic** • London, England
Diploma in Home Economics and Catering (1978 Graduate)
• Certificate in Cookery for the Catering Industry (awarded by The Hotel Catering Institutional Management Association of London); Certificate in Cooks Professional (awarded by the National Council of Home Economics, London)

PROFESSIONAL CERTIFICATIONS

• **The Educational Foundation of the National Restaurant Association**
FMP — Food Service Management Professional (1994)
• **Cornell University** • Ithaca, NY
Professional Development Program — Front Office Management Certification — Concierge Management Certification — Food Service Management Certification
• **Certified Tips Trainer — Certified Food Service Sanitation Trainer**
• Successfully completed numerous continuing professional education seminars conducted by Hilton Hotel Corporation and Marriott Hotel Corporation (including Priority One Guest Satisfaction Program, Performance for Excellence, Yes I Can)

AFFILIATIONS

• **American Hotel and Motel Association**   • **Global Hoteliers Club**
• **National Restaurant Association**

SANDY SUSTACHEK
151 W. Passaic Street
Rochelle Park, New Jersey 07662
(201) 555-3772

Detailed portfolio of experience supported by a solid work history. Minimization of weak education by promoting Major Projects and Accomplishments on first page and listing the education at the end of the document on the second page.

HUMAN RESOURCES & TRAINING PROFESSIONAL

- Strong HR generalist experience with a proven talent in the development and implementation of training programs for exempt and non-exempt personnel; proven ability to develop material, impart knowledge and update programs as needed.
- Strong research and analytical abilities; notable experience in the management and reduction of costs related to liability and insurance.
- Recruiting activities included all aspects of screening, interviewing, hiring and orientation for union and non-union staff.
- Continuously updates knowledge relevant to workers' compensation, ADA, EEO, Family Medical Leave Act, OSHA, D.O.T., etc; develop and implement new procedures to ensure compliance.
- Benefits administration experience includes program development, maintenance of costs through negotiations and the development/implementation of alternative benefit programs.
- Assets in collective bargaining activities; successfully administers benefits to assist in successful negotiation.
- Generates on going bottom line savings through the introduction of various cost cutting programs.

MAJOR PROJECTS & ACCOMPLISHMENTS:

♦ **Reduced corporate liability insurance costs by 1/3; restructured liability insurance management program to facilitate savings.**

♦ **Reduced annual costs by $250K by establishing an effective self-insured plan for supplemental disability.**

♦ **Facilitated a 1/3 reduction in medical benefit costs by implementing an HMO as an integral part of benefits plan.**

♦ **Reduced annual paid claims by 45% by developing and implementing aggressive claims handling procedures.**

♦ **Increased parent company revenues by $185K in 6 months by re-instituting billing of subsidiary companies for healthcare coverage.**

♦ **Played an integral role in 12 different collective bargaining agreements - assisted the Senior V.P. by providing benefit guidance.**

♦ **Attained a 50% conversion from indemnity to managed healthcare through the creation and implementation of a successful managed care program.**

(continued...)

SANDY SUSTACHEK   -Page Two-

(201) 555-3772

PROFESSIONAL EXPERIENCE:

GLATT AIR TECHNIQUES - Ramsey, New Jersey   1992 - Present

Human Resources Manager • 1994-Present
Direct staff in daily HR operations including employee recruitment, new hire orientation, coordination of COBRA, administration of Federal regulations, collective bargaining, benefits administration and procedure development. Maintain OSHA recordings. Coordinate payroll for 150 employees, interfacing with key intra-department personnel. Provide assistance with employee problems and serve as the first line in resolving issues. Team member of the Safety Committee. Revised Company Handbook and implemented organizational policy. Attended comprehensive ADP training program.

Administrative Director/Office Manager • 1992-1994
Managed all office activities - created and administered office procedures. Implemented new accounting software; handled all training.

PILGRIM GROUP, INC. - Fort Lee, New Jersey   1986-1992

Manager of Special Projects
Trained and supervised three staff members in various interdepartmental projects, preparation of monthly trade reports and purchasing activities.

EDUCATION:

Union College - Cranford, New Jersey
A.A.S. in Human Services

Additional Training:
Society of Human Resource Management:
Courses in Employment Law, Disability, Workers' Compensation

43—INDUSTRIAL ENGINEER

TIMOTHY P. O'CONNOR

15 North Mill Street • Nyack, New York 10960 • (914) ...

PROFILE

A results-oriented *Industrial Engineer* with proven abilities in improving efficiency of operations, developing effective organizational/flow charts and detailing project information to determine effective functions for manufacturing operations. Able to identify areas of duplication and provide recommendations resulting in increases in productivity and profitability. Demonstrated ability to motivate staff to maximum productivity and control costs through the most effective uses of manpower and available resources.

AREAS OF EXPERTISE

Program Development • Streamlining Operations • Forecasting
Project Restructuring • Costing • Labor Utilization Analysis

EXPERIENCE

Astrosystems Inc. - Lake Success, New York   1966/Present
Industrial Engineer/Production Manager • 1990/Present

- Manage and coordinate production operations for this $30 million manufacturer of computer peripherals, communication equipment, high voltage power supplies, automatic test equipment, position control electronics, electromechanical products and other related products for military and industrial customers.
- Schedule and monitor manufacturing, distribution, handling, shipping and customer follow-through.
- Calculate materials requirements and communicate with purchasing personnel to ensure timely delivery of raw materials/components in support of manufacturing schedules.
- Control quality and costs through strict attention to detail, compliance with specifications and selective purchasing/inventory management.
- Supervise a staff of 80 and communicate directly with the Vice President in Charge of Production regarding all pertinent issues and concerns.

Accomplishments

- *Provided valuable input and guidance in the design and development of various high voltage linear and switching power supplies.*
- *Devised and implemented plans and processes to facilitate goal attainment.*
- *Representative projects include:*
    - *Litton - TAOM: Tactical Air Operation Module and MCE: Module Control Equipment*
    - *U.S. Air Force - Cordwood-ANCU/F14 Aircraft and VHF Relay Transmitter*
    - *U.S. Navy - V.A.S.T. AN/UMS-247*
    - *Raytheon - DRFM: Digital FR Memory Unit*

TIMOTHY P. O'CONNOR • *PAGE TWO*

Professional Experience Continued...

General Manager • *1980/1990*
- Managed all day-to-day administrative and technical activities in a company owned, alternate manufacturing facility.
- Directly supervised design, manufacturing and distribution of industrial and military electronics.
- Provided accounting department personnel with accurate documentation to simplify billing/invoicing.

Technical Supervisor/Technician • *1966/1980*
- Participated in all phases of design and manufacturing of toroidal transformers, A to D/D to A converters, linear and switching power supplies.
- Tested and debugged electronics throughout all stages of development.
- Coordinated and managed projects from prototype to final assembly.
- Applied extensive expertise in computer peripherals, communications equipment, displays and transducers for military and industrial application.

EDUCATION • AFFILIATION

- **New York Institute of Technology** - *Associate of Applied Science in Engineering*
- **RCA Institute** - *Associate of Applied Science in Industrial Engineering*
- **I.E.E.E.** • *Active Member*

REFERENCES WILL BE FURNISHED UPON REQUEST

Executive-level resume with power box to open the resume.

Areas of Strength section indicates contributional skills and value followed by a strong section listing computer skills.

Bullets accentuate key responsibilities and accomplishments.

MARGARET BREDEN

Address
6080 Salsbury Lane
Boynton Beach, FL 33437
Home: (561) 555-5580
Fax: (561) 555-4590

INFORMATION TECHNOLOGY PROFESSIONAL
Systems Design / Product Engineer
Strategic Business Planning / Senior-Level Project Management

Expert in the design, development, and delivery of cost-effective, high-performance technology solutions to meet challenging business demands for well recognized international corporations including Motorola and Hewlett Packard. Extensive qualifications in all facets of project lifecycle development - from initial feasibility analysis and conceptual design through documentation, implementation, and user training/enhancement.

Equally effective organizational leadership, team-building, and project management experience - introducing out-of-the-box thinking and problem-solving analysis to improve processes, systems, and methodologies currently in place to exceed business goals and to perpetually delight shareholders and customers.

AREAS OF STRENGTH

* National and international marketing
* Business research and analysis skills
* Strategy identification and implementation skills
* Personnel training, team-building, and supervision
* Customer service and retention management

* Supply chain management
* Cost reduction methodologies
* Presentation and public speaking skills
* Finance management/project budgeting
* Quality control management

COMPUTER / TECHNOLOGY

Excel, Word, PowerPoint; Programming Languages including Basic, Assembly, C; Use of Unix Based, PC or Macintosh Platforms; RF, Wireless, Microcontroller/Microprocessor System Skills; Digital and Analog Hardware Experience; IC Design Tools: VHDL, Spice, Cadence, Synopsis; Knowledge of technology Required to Implement Small, Portable, battery-Powered, Wireless Multimedia, Consumer Products into the Future.

PROFESSIONAL EXPERIENCE

MOTOROLA, Boynton Beach, FL   1990 - Present
STAFF ENGINEER (1996-Present)
* Identify technology to meet future needs in the areas of wireless communication products for CTSO division (Core Technologies Systems Organization).
* All program direct material costs tracked. Perform comparisons of plans versus actual. Track key product performance criteria. Produce competitive analysis reports of our products versus the competition in terms of performance, direct material cost, and average selling price.
* Forecast systems, discrete and integrated semiconductor technology trends.
* Benchmarking of CMOS, BiCMOS, and SOI semiconductor technologies.
* Work closely with market visionaries to identify future product features - especially those needed for future system integration to be used as a driving force for early technology identification.
* Provide strategic insights - marketing, operational, and product development/enhancement - in making recommendations on what business we should do, with whom, for how long, and why to meet business objectives (time to market, features, and related costs).
* Vendor liaison, i.e. seeking out needed technology, reviewing technology roadmaps, and managing the non-disclosure agreements for the division, projecting base cost of all semiconductor chips to be designed. Manage Non-Recurring Engineering (NRE) and Request for Quote (RFQ) processes for the division.
* Memory statistics and benchmarking of embedded/external NVM, ROM, and RAM in terms of relative cost, die area consumption, relative cost trends over time, and cost trend for various memory sizes.
* Compiled PPG's roadmap for 1998-2000, and identified technology needed for its realization.

120

Margaret Breden
Page 2

PROFESSIONAL EXPERIENCE (Continued)

- Responsible for implementing cost reduction methodologies to reduce direct material cost without adversely affecting manufacturing cost through understanding the supply chain for each technology.
- Perform analysis of different system approaches to meet market needs, i.e. time to market, features, and costs.
- Received professional recognition as an "Intersector and International Resource."

SENIOR ELECTRICAL ENGINEER (1990-96)
- Changed way of thinking as an option, not previously explored, to meet market needs that is a cost analysis of outsourcing versus traditional in-house methods. This set a precedent that is being used more and more as a strategy to meet business goals and objectives.
- Program management skills including product feasibility study, specification and contract development, scheduling project tasks, managing outsourcing activity (including legal implications), advertising, and customer support.
- Digital and analog design simulation; system level simulation and verification; interfacing with layout design personnel; and evaluation/debugging of integrated circuit.
- Analytical probing including utilization of laser and FIB technology.
- Provide technical support to customers to enable the implementation of power management and device driver integrated circuits.
- Provide documentation/transfer design to outsourcing company for continued manufacturing and technical support.

HEWLETT PACKARD, Boynton Beach, FL   1988 - 1990
ELECTRICAL ENGINEER (1990)
INTERNSHIP (1988-89)
- Developed analog hardware used in hospital diagnostic patient equipment.
- Designed, developed, and tested a complete software package that included on-line help and manual - released August, 1989. Software provided a mouse-driven environment in which the user could program a digital oscilloscope without knowledge of the oscilloscope programming language or IEEE-488 mnemonics.
- Developed logic analyzer preprocessor module for a graphics coprocessor. Packaged included a digital controlled logic hardware probing adapter and inverse assembly software.

EDUCATION/TRAINING

CITY COLLEGE OF THE CITY UNIVERSITY OF NEW YORK, New York, NY
Bachelor of Engineering in Electrical Engineering (BSEE), 1990

UNIVERSITY OF MIAMI, Miami, FL
MBA: International Business, 12/97

Seminars/Workshops/Continuing Education
* Anglesaria Method of Cost Reduction
* The Changing Structure of the Semiconductor Industry
* Technology and Alliances in the Economy
* Issues and Opportunities in the Next Recovery

PROFESSIONAL AFFILIATIONS & COMMUNITY INVOLVEMENT

Member: Society of Women Engineers
Member: Institute of Electronic and Electrical Engineers
Mentor: Underachieving Middle School Students

References & Supplemental Information Furnished Upon Request -

45—INSURANCE SALES

Executive-level, two-page resume with extensive information including thorough listing of education, training, and professional affiliations.
Employment section packed with key accomplishments and value-oriented information.
Lots of numbers, figures, and percentages to support the accomplishments.

ROY WHEELER

13 WALNUT STREET, PITTSBURGH, PA 15012   **412.555.0847**

INSURANCE SALES PROFESSIONAL. TOP PERFORMER.

B.S.B.A. / Financial Planning, Insurance and Investment Sales / Increased Customer Base
Pioneered Efforts In New Territories / Innovative In Marketing / Multi-Million Dollar Producer

- 13-year track record for success in networking and selling - - #1 Sales Rep consistently. 8 years in insurance, 5 in retail merchandising/distribution.

- Took initiative in self-marketing and self-promotion. Developed promotional materials, grew territories by up to 45 percent, achieved a 100 percent client retention rate and generated strong referral networks.

- Capably positioned the companies represented as a preferred provider through extensive personal contact and a mutual respect for clients' time.

- Special knowledge and continuing education in investments, retirement and estate planning, securities, mutual funds, annuities, and life and health insurance.

- 5-year entrepreneurial background provides a "whatever takes" commitment to mutual success. Created winning campaigns that boosted sales an average of 22 percent annually. Controlled expenses, motivated staff and negotiated competitive terms saving 35 percent.

SALES AND MANAGEMENT EXPERIENCE

A VERY LARGE INSURANCE COMPANY, SOMEWHERE IN, PA   **1992 to Present**
Direct Client Contact - Financial Sales and Marketing - Top Performer - 600+ New Accounts
REGISTERED REPRESENTATIVE - PENNSYLVANIA AND WEST VIRGINIA
Successfully convert new investors into educated clients through persistence, patience, keen listening, and the ability to establish trust. Added $15 Million in new business. Farm and secure leads in the most productive, cost-effective means possible.

- Initially acquired clients exclusively via cold-calling. Achieved increased productivity by hiring outside telemarketers. Warm-call qualified individuals, coordinate appointments and ask for their business.

- Landed 600+ new accounts throughout PA and WV region over a 4-year period, representing more than $15 Million in new business.

- Realize a 98 percent client retention rate - - by keeping in touch via personal contact and a self-developed newsletter discussing investment tips, generated increased networking opportunities, a more educated clientele, and better quality attention to their priorities.

- Market annuities, securities, mutual funds, estate planning and personal retirement services, life and health insurance to personal and corporate investors.

- Through interactive conversation, identify clients' long- and short-term needs. Stress alternatives in their best interest - - in one instance, saved an investor over $110K in inheritance taxes.

- Recognized for outstanding performance. Sales Agent of the Month 15 times over a 3-year period. Sales Agent of the Year in 1996 and 1997. Won various in-house contests.

BEST RATE INSURANCE, PITTSBURGH, PA   **1990 to 1992**
Independent Agent - Regional Firm - Auto, Life and Health Insurance - Top Representative
REGISTERED REPRESENTATIVE - PENNSYLVANIA AND MARYLAND
Generated $1.5 Million in new business within a 22-month period. Prospected new customers via targeted direct mailings. Became involved in local events via memberships with various community organizations.

- Stepped into a 9-county, 2-state territory and aggressively pursued leads. Added 200 new accounts, earned membership into "Peak Achiever's Club" in 1991 and 1992.

- Using continued mailings and follow-up surveys, acquired better qualified leads - - closed 8 out of 10 calls. #1 representative of 24.

ROY WHEELER
INSURANCE SALES PROFESSIONAL
2 of 2

SALES AND MANAGEMENT EXPERIENCE (cont.)

BLUE NOTE'S, ANYVILLE, PA   **1985 to 1990**

MANAGING PARTNER

Rebuilt business, defined a winning marketing strategy that enabled continued growth, and negotiated a profitable final sale. Oversaw all operations-related priorities, from personnel and service to advertising, accounting, payroll, purchasing and taxes.

- Took over a failing restaurant/bar business, redefined it as an upscale jazz club. Sought and attracted local talent for live performances 6 nights a week.

- Hired, trained, disciplined, scheduled and motivated a staff of 12 wait/kitchen staff and an assistant manager. Calculated payroll, managed inventory, and worked in all areas of the operation as needed.

- Reevaluated the vendor network, reconstructed accounts, and negotiated more competitive terms resulting in a savings of 34 percent and improved product quality and delivery terms.

- Solely handled accounting and taxes, managed A/P and A/R, and monitored all financial aspects using QuickBooks 4.0 for Windows 95.

- Lead decision maker for advertising - - liaison to media representatives. Handled all creative copywriting, secured free publicity and coverage with 3 local magazines and 4 radio stations. Created an ongoing campaign that boosted sales an average of 22 percent annually.

GREENER GARDENS, FRAMINGHAM, PA   **1980 to 1985**

SALES REPRESENTATIVE - OHIO REGION

Added 100+ new accounts, representing a 45 percent increase in business. Developed own newsletter, "The Green Leaf," discussing gardening tips and statistics that greatly enhanced customer communication while generating referrals for a national distributor of Miracle and Ortho Products.

- Landed an account with a 15-location regional greenhouse franchise - - only sales rep in company history to break through the hierarchy and deliver the order.

- Pioneered efforts in a new territory - - increased sales 45 percent over an 11-county area via a self-developed promotional newsletter.

- Implemented corporate planograms and negotiated for prime merchandising space at 200+ locations. Serviced accounts and maintained personal relations with managers for major stores such as Kmart, Wal-Mart, Lowe's and Stambaugh's, as well as smaller independent retailers.

EDUCATION, LICENSURE

AMERICAN COLLEGE - ROBERT MORRIS COLLEGE, PITTSBURGH, PA
Financial Planning Courses (1994 to 1995)

CALIFORNIA UNIVERSITY OF PENNSYLVANIA, CALIFORNIA, PA
B.S. In Business Administration / Concentration: Marketing

CONTINUING EDUCATION ... Graduate, Dale Carnegie Course. Sales, Marketing And Motivational Seminars. Corporate-Sponsored Personal Retirement And Estate Planning (1,2) Courses.

INSURANCE, SECURITIES AND INVESTMENTS LICENSES ... Life And Health Insurance. NASD Series 6 Mutual Funds And Variable Annuities. Series 63 Pennsylvania Securities.

PROFESSIONAL AFFILIATIONS

Member/Board Member/Treasurer - Pittsburgh Industrial Development Association. Board Member - Pittsburgh United Way. Board Member/Committee Chairperson - Mon Valley Progress Council. Board Member - Pittsburgh Area Revitalization Corporation. Member/ Treasurer/Vice President - Monongalia Neighborhood Boys Club. Member/ Board Member/President - Monongalia Area Chamber of Commerce. Member/Board Member/President - Monongalia Rotary Club. Board Member - PA Economy League, Allegheny County.

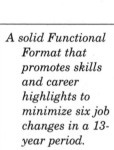

Diana A. Jacob

172 River St., Douglaston, NY 11363 ➤ 516-555-4201 ➤ DAJ@aol.com

Residential and Showroom Interior Designer

Over ten years of experience in the planning and coordination of $10 million, 55,000+ square foot fine furniture showrooms for retailers including Macy's Furniture Store and The Place Furniture Galleries.

Plan and implement total showroom design from inception to grand opening. Showrooms have been recognized by industry representatives as some of the best in the tri-state area.

Manage complete customer design assignments for Macy's, Ethan Allen, Classic Galleries, Georgetown Manor and Levitz, in home, office or retail environments.

Expert in trend analysis, merchandising and management. Highly creative, capable and motivated; readily inspire confidence of clients.

Areas of Expertise

- ➤ residential / commercial design
- ➤ retail showroom design
- ➤ home furnishings trend analysis
- ➤ floor plans / space planning
- ➤ construction planning
- ➤ vignette / color coordination

- ➤ manufacturers' representative interface
- ➤ commercial shoot consulting
- ➤ accessories purchasing / placement
- ➤ faux painting / stenciling
- ➤ themes / holiday displays
- ➤ staff management / budget oversight

Career Highlights

Managed design studio as in-store designer for Macy's 45,000 square foot Farmingdale Furniture Store's showroom and customers. Worked with high expectation clientele and a range of fine furniture, accessories and floor coverings. Developed a cohesive and distinctive showroom that became one of Long Island's premier locations for fine furniture.

Totally re-designed and opened The Place Furniture Galleries' new 55,000 square foot furniture showroom in Farmingdale Long Island. Independently planned space for grouping of galleries, drafted floor plans, and interfaced with store management and buyers. Purchased accessories to compliment vignettes and created elegant look to enhance moderate furniture collections.

Designed The Place Furniture Galleries' new children's area, leather area, and motion area. Determined design strategy, colors, furniture placement and accessory selections.

Achieved second place award in eleven store display competition; participated in interior design showcases. As Nassau Coliseum Home Show designer, developed and set-up 20'x40' vignette.

A solid Functional Format that promotes skills and career highlights to minimize six job changes in a 13-year period.

Emphasis of the resume is on what she did, not where she did it.

Graphics are used to underscore creativity.

Diana A. Jacob   *2*

Professional skills

➤ Conceive and draft floor plans and designs for homes, offices and commercial space. Determine all components of interiors—color schemes, furnishings, lighting, accessories, window treatments, etc. Emphasize client needs assessment; relate well to all types of people.

➤ Develop and implement retail client design service programs. Determine client needs; draft plans and create design proposals. Recommend merchandise necessary for fulfillment of goals. Provide home visits, in-store consultations, and coordinate finalization of projects.

➤ Plan and arrange complete furniture store room settings. Make quick and intuitive decisions for placement and floor coordination of unexpected stock arrivals. Redo vignettes and accessories on a daily basis in response to new shipments and constant rotation of furniture.

➤ Buy accessories and work with a semi-annual budget. Supervise furniture store operations and handle customer service, sales supervision, opening and closing of building, security, cash vault. Sell on floor when needed.

Experience

Interior Designer, Showroom Designer, Accessories Buyer   1995 to present
The Place Furniture Galleries, Farmingdale, NY

Interior Designer, Design Studio Manager   1993 to 1995
Macy's Furniture Store, Farmingdale, NY

Interior Designer, Showroom Designer   1991 to 1993
Levitz Furniture, Garden City, NY

Interior Designer, Sales   1990 to 1991
Ethan Allen, Hicksville, NY

Interior Designer, Sales   1989 to 1990
Classic Galleries, Huntington, NY

Interior Designer, Sales   1986 to 1989
Georgetown Manor, Westbury, NY

Education

Interior Design Certificate   1986
Metropolitan Institute of Interior Design, Plainview, NY

A.A.S. in Design   1975
Fashion Institute of Technology, New York, NY

Internet-friendly and scannable resume void of lines, accentuations, and graphics. Resume has keywords throughout the document indicating proficiency in his field. Highlights section effectively addresses his accomplishments.

PATRICK D. DUDASH
Phone: 214~555~0300

2820 HILLCREST, SUITE 309   DALLAS, TX 75200

INTERNATIONAL AFFAIRS — PROJECT DIRECTOR

Dedicated career professional. Top-level experience in international project development. Practiced in portfolio management, project proposals and budget formulation. Highly effective negotiator and interface with appointees and government officials at all levels—including the President of the United States. Played a key role in effecting ground-breaking results that escalated free and fair election processes for emerging democracies in Latin America. Brings two Bachelors of Arts degrees. Widely traveled, **fluent Spanish**—instructor level.

PROFESSIONAL EXPERIENCE

NATIONAL DEMOCRATIC INSTITUTE FOR INTERNATIONAL AFFAIRS (NDI)
WASHINGTON, D.C. ~ 5/93 TO 4/98
Non-profit organization working to strengthen and expand democracy worldwide by providing assistance to civic and political leaders advancing democratic values, practices and institutions.

PROGRAM OFFICER/SPECIAL ASSISTANT TO THE DIRECTOR OF LATIN AMERICA ~ 12/95 TO 4/98

Acted on Director's behalf in Director's absence. General management of staff (11) and $1.75M International Political Development Project Portfolio of Latin America and the Caribbean. Generated funding from U.S. and international clients by conducting back-to-back project proposal presentations when in town. Coordinated and directed programs in Latin American countries that dramatically increased voter awareness and rallied support for free and open elections.

- Led NDI Assessment Missions to Latin American countries.
  - Provided Director's office with in-depth analysis of each country's political and socio-economic environment.
  - Implemented and directed all planning and administrative components of Voter Education and Pollwatcher Training programs in Guatemala.
- Coordinated 9-country Regional Anti-Corruption Program in Argentina in cooperation with Transparency International.

FIELD DIRECTOR, DOMINICAN REPUBLIC
CARTER CENTER OF EMORY UNIVERSITY/NDI ~ 1-7/96

PROJECT PROFILE:
International pressure following the flawed 1994 Presidential election in the Dominican Republic called for early elections to be held in 1996. For these elections, NDI/Carter Center—actively involved in the Dominican Republic electoral process since 1990—organized a group of international observers to effectuate true reform in a country plagued by election fraud, political unrest and violence during electoral processes.

Directed two Preelection Assessment Missions and two 45-member International Observer Delegations to drive Carter Center/NDI initiative for the 1996 Dominican Republic general election—led by former President Jimmy Carter. Reported to Carter Center/NDI on post-election environment.

- Primary point- of-contact for Dominican Republic political parties, civic organizations, United States Embassy, and Dominican Election Commission.
- Chief liaison between President Carter and President Joaquin Balaguer.

PATRICK D. DUDASH   PAGE 2

EDUCATION

BACHELOR OF ARTS DEGREE IN INTERCULTURAL COMMUNICATION
BACHELOR OF ARTS DEGREE IN SPANISH LITERATURE
UNIVERSITY OF OKLAHOMA~ NORMAN, OKLAHOMA ~ MAY, 1993

- President, University Study Abroad Alumni Association
- Corresponding Secretary, Fraternity of Phi Gamma Delta
- Chair, International Advisory Council
- Member, International Business Association
- Public Relations Officer, OU Intrafraternity Council

JOINT BACHELOR OF ARTS DEGREE IN SPANISH LITERATURE
UNIVERSIDAD DE LAS AMERICAS ~ PUEBLA, MEXICO

CAREER DEVELOPMENT

NATIONAL DEMOCRATIC INSTITUTE FOR INTERNATIONAL AFFAIRS (NDI)
PROGRAM ASSISTANT ~ 1994 TO 1995

Implemented local Government training program in El Salvador. Conducted needs assessments. Provided organizational and communications advice and training to in-country partner organizations.
International Election Observer:
- 1994 - Dominican Republic
- 1994 - Mexico
- 1995 - Guatemala

PROGRAM INTERN ~ 93-94

- Supported implementation and management of Democratic Development Programs in 8 Latin American countries.
- Organized multi-city Election/Campaign Education Program for 8 Russian political dignitaries during 1994 United States general elections.

AMERICAN FIELD SERVICE INTERCULTURAL PROGRAMS (AFS)
Non-profit organization that conducts international exchanges for high school student to gain the knowledge, practice the skills, and acquire the attitudes needed to live and work in a multicultural, global society.

VOLUNTEER ~ 1987 TO PRESENT
- Donate time and expertise to Student Sending Center for Dallas/Fort Worth Metroplex
- Student Sending Director of National Capital Area, Washington, D.C. - 1994 to 1995
- AFS Intercultural Student Liaison Director, Oklahoma - 1988 to 1993
- Activities Director for 600-student AFS Conference, Dallas, Texas - 6/90
- Student Participant, Dominican Republic - 1987 to 1988

James A. Trigali

36 Mary Avenue • Portland, Oregon 97201

> *The shaded box of testimonials is a unique approach that will work if they are complimentary and from key contacts.*

> "Jim's leadership, organizational skills, and initiative have been exemplary, evidencing his consistent commitment to excellence and strong customer service results. It has been a pleasure and privilege to work with him." (*Supervisor, 8/93 to 5/95*)

> "Jim provides an unassuming but very supportive and strong style of leadership that not only team members but members of the whole unit respect and value. He is a true team player." (*Manager, 5/94 to 9/94*)

> "Jim Trigali is definitely an employee that any manager would welcome—he is supportive of management ideas but has the confidence and strength to raise critical concerns that only make the organization a better one." (*Manager, 5/94 to 9/94*)

HIGHLIGHTS
- Comprehensive knowledge of defined contribution 401(K) pension concepts and regulations
- Effective leadership and team-building skills
- Extremely well organized
- Proven customer service skills
- Excellent PC skills, including Windows, WordPerfect, and defined contribution 401(K) daily processing, stock, and mutual fund system logic

EXPERIENCE

CIGNA RETIREMENT & INVESTMENT SERVICES, Portland, OR
Plan Analyst I-IV (1990 to 1996)
Processed all contributions, loan repayments, benefit disbursements, fund transfers, and nonfinancial changes for daily/periodic 401(K) pension record-keeping transactions, including cash, mutual fund, stock, and guaranteed investment contract money movements. Resolved client questions and problems. Interfaced productively with team, department, field, and sales personnel. Completed, verified, and mailed all required client reports, plan year packages, and auditors' requests. Processed ADPs/ACPs and correction of year-end participant tax information.

Achievements:
- Promoted four times in four years
- Consistently met or exceeded all established standards for timing and accuracy
- Selected as one of first two people to participate in changeover to a daily 401(K) processing environment from a monthly processing environment, and quickly adapted to new system
- Chosen to train new hires and other plan analysts transitioning to the daily processing environment
- Served as technical liaison with systems personnel in identifying, communicating, and resolving processing problems with the new system
- Served effectively as team leader and account manager for new *pilot team* concept while maintaining own book of business

AWARDS

PRESIDENT'S CLUB AWARD (1994)
Awarded for contribution to the development of the defined contribution 401(K) daily processing record-keeping system.

EDUCATION

PACIFIC NORTHERN COLLEGE, Portland, OR
Bachelor of Science, Business Administration (1990)

49—JOURNALIST

MICHAEL R. BARTLE
151 W. Passaic Street
Rochelle Park, New Jersey 07662
Tel: (201) 555-3772   Email: xxxxxx@mindspring.com

EXPERIENCED JOURNALIST

A proven professional with experience in covering a variety of topics including general assignments, business reporting, courthouse reporting, special projects, human interest stories and municipal beat coverage. Versed in researching topics and developing relationships with key sources. Accustomed to demanding deadlines and high-pressure situations. Experienced in navigating the Internet to research daily stories. Versed in Windows and Word. Freelance for the Associated Press. Recipient of numerous industry honors and awards for outstanding reporting and innovative journalism.

PROFESSIONAL EXPERIENCE:

THE NORTH JERSEY HERALD NEWS	Passaic, New Jersey
Reporter	1992 - Present

Accountable for generating daily and weekend story ideas for a daily newspaper serving Passaic, Bergen and Essex counties with a circulation of 55,000+; research topics and foster positive working relationships with sources.
Gather and coordinate photos for stories.
Provided coverage of Passaic County and Bergen County news from courthouse bureaus.
Instrumental in generating stories with business reporter.

Highlights:

Authored a series involving coverage of a convict on home detention who murdered a Paterson teenager: stories prompted the State to suspend its electronic anklet monitoring program.
Covered President Bush's visit to northern New Jersey, and its impact on local residents.
Generated coverage of Christie Whitman's gubernatorial campaigns in both 1997 and 1993.
Covered the Pope's visit to Northern New Jersey.
Covered major news events including an explosion which killed 5 people, a post office shooting which killed 4, and the Paterson riots which took place in 1995.
Authored stories regarding "Megan's Law" - from inception in 1995 to reporting on current cases.
Reported extensively on NJ State's sex offender treatment center, AIDS and gay issues.
Authored a Lifestyle story which was picked up and broadcast by Japanese television stations.
Conducted in-depth interviews with politicians, sports stars and celebrities including presidential candidate Ross Perot, Sinn Fein leader Gerry Adams, Jets football star Dennis Byrd, Christopher Reeve's wife (Dana Reeve) following his tragic accident and subsequent paralyzation, and "Grease" star, Olivia Newton-John.

THE DAILY JOURNAL	Elizabeth, New Jersey
Reporter	1988 - 1992

Generated daily and weekend story ideas for a 30,000+ circulation newspaper covering Union County, NJ.
Assigned to cover two high-profile cities, Linden and Rahway, New Jersey.

EDUCATION:

Bachelor of Arts in Journalism/News Editorial
Concentration in Broadcast Journalism   Minor in Speech Communications
Shippensburg University, Shippensburg, Pennsylvania

Writing samples and portfolio available upon request

50—LANDSCAPE DESIGN

ROBERT BYRNES, ASLA —

Telephone / Voice Mail: 214-555-2

LANDSCAPE DESIGN —

Design, install and maintain noteworthy landscapes in Dallas and closely surrounding communities—a showplace market that includes the legendary azaleas of Turtle Creek. Artistic and technical site planning expert. Ensure coordination and integration of architecture and engineering work scopes. Preferred seasonal color designer for the Park Cities, North Dallas and other exclusive interior communities. Work published in *Southern Living* and *D Magazine*.

PROFESSIONAL EXPERIENCE —

SEASONS — LANDSCAPE DESIGN & LAWN CARE
PRINCIPAL

DALLAS, TEXAS
1994 TO PRESENT

Contracts and manages all aspects of the firm's projects from design to completion.

SELECTED RESIDENTIAL PROJECTS —

411 Beverly Drive - $7M executive estate on Turtle Creek:
- Steeply sloping site focuses on flagstone sun terrace, spa/pool and low walls of cascading azaleas.
- Water features include lily pond, spa-to-pool waterfall and front entry fountain.

4417 Swiss Avenue - Dallas Historic Homes Reclamation Project:

Main thrust of landscape design was to leave established tree canopy intact while creating sunny, flat lawns for play and entertaining. Installed privacy screen planting at the street, perimeter fence and gate.

SELECTED COMMERCIAL / RECREATIONAL PROJECTS —

Addison Community Arts Center:

Created softscape design and developed site plan. Laid out circulation areas and provided unique design alternatives for an outdoor courtyard.

Bedford's Stoneleigh Park:

Designed new playground site of sand filled terraces wrapped by a stone bench wall that protects adjacent trees.

PROFESSIONAL REGISTRATION —

1993 - Texas Registered Landscape Architect #52973

PROFESSIONAL AFFILIATIONS —

American Society of Landscape Architects
Park Cities Design Review Board - 1994-98
Texas 295 Corridor Study Committee - 1994-97

EDUCATION —

BACHELOR OF SCIENCE - LANDSCAPE ARCHITECTURE - 1993
TEXAS A & M UNIVERSITY - COLLEGE STATION, TEXAS

51—LAW CLERK/E-MAIL RESUME

ELEANOR DUFFEK
2281 Dallas Parkway
Alhunter, TX 75240
(972) 555-9282
eduffek@yahoo.net

> *Simple layout and font designed for E-mail and transport via the Internet.*
> *Simple font that is monospaced.*
> *No elaborate formatting, bold, underlined, or italicized fonts.*

LAW CLERK

OVERVIEW: Skilled law professional with experience in writing
legal briefs, drafting legal letters, creating affidavits and
prescreening clients.

STRENGTHS:
Computer Skills:
Word Perfect 5.1/6.0
Windows Applications
Microsoft Word
Excel
Super Writer

Personal:
Work well (and competently) under stress
Present a very professional demeanor and strong corporate image
Work in harmony with all staff; a team-spirited professional
High energy, motivated, and dependable
Excellent communication skills and problem solving skills
Trustworthy, ethical. Recognized for high degree of integrity

PROFESSIONAL EXPERIENCE:

Smith and Smith, Esq., Westlake, TX   1989 - Present
Legal Assistant to law firm partner

-Draft general legal contracts for partner review
-Prepare client and witness statements
-File motions at county court
-Assist and take leadership role in processing discovery,
including interrogatories, statements and affidavits

EDUCATION:
University of Texas-Dallas, Dallas, Texas
Bachelor of Science, Criminal Justice

INTERESTS:
Enjoy: Exercise, Reading, and Outdoor Activities

52—LAW ENFORCEMENT

Lester Hingis

10 Badview, Apt. 36 • San Antonio, Texas 78213 • (210) 555-9999

Summary of Qualifications

Ten years of distinguished military service in law enforcement, including supervisory positions. Additional experience as a surveillance camera operator for a St. Louis-based casino company. Certified Customs Officer with additional training in field operations and security activities for high-profile military functions. Hold a Top Secret SSBI clearance. 9mm handgun and M-16 rifle qualified. Strong knowledge of general police duties including:

Investigation/Search Procedures	Evidence Collection/Protection	Resource/Protection Programs
Traffic Control/Operations	Security/Emergency Response	Defense Management
Dispatching	Apprehension/Advisement of Rights	Testifying in Court

Professional Experience

PRESIDENT CASINO ON THE ADMIRAL, 1998-Present

SURVEILLANCE OPERATOR: Develop and maintain a closed circuit TV surveillance system within the guidelines and regulations set forth by the Gaming Commission. Protect company assets. Ensure adherence to systems of internal control and procedure. Detect cheating and other criminal activity in full cooperation with the Commission in an effort to maintain the integrity of the company.

UNITED STATES AIR FORCE, 1988-1998

ELITE GUARD SUPERVISOR (1993-1998): Supervised four senior airmen responsible for control entry of three centralized headquarters. Led security response teams in support of a unified command, major command, and a field operating agency. Perform personal security for distinguished visitors and key personnel. Screened and processed more than 5000 workers and visitors daily. Provided customer service and headquarters-related information to visiting military and civilian personnel.

- Performed major military ceremonies including visits of the Chief of Staff, French Air Force, and Secretary of Defense.
- Selected to train others on a new automated entry control and alarm system.
- Competitively selected to provide ceremonial support during the visit of the Secretary of the Air Force; earned an Air Force Commendation Medal, a Certificate of Appreciation from the security police commander and numerous individual commendations for outstanding job performance.

LAW ENFORCEMENT SPECIALIST (1988-1992): Investigated offenses involving vehicular and pedestrian traffic. Protected personnel, equipment, and facilities. Monitored radio communications and provided routine vehicle patrol. Performed operator maintenance on assigned equipment, materials, and vehicles. Apprehended and detained suspects; accepted custody of military personnel apprehended by other agencies. Dispatched law enforcement and security patrols.

- Ranked in the top 2% among peers; hand-picked by specific VIP military officials to lead anti-terrorist and security activities at an elite post in Germany. Earned a commendation medal and numerous letters of appreciation from traveling dignitaries.
- Provided personal security for General Colin Powell during a post-Gulf War visit to Iraq.
- Provided convoy security over unsecured roads to ensure successful delivery of critical resources during a vital humanitarian relief effort in the Middle East. Earned an Air Force Achievement medal.

Education / Specialized Training

- B.A. Degree, Management/Communications, Concordia University Wisconsin, Mequon, Wisconsin, 1987
- A.A. Degree Program (61 credit hours), Criminal Justice, Community College of the Air Force, 1983
- Antiterrorist Driving and Surveillance Detection Course, International Training Incorporated
- Distinguished Graduate, Air Base Ground Defense Training, Fort Dix, New Jersey

53—MBA GRADUATE/STUDENT

MICHAEL BOUCHER

123 44TH STREET, PITTSTOWN, PA

745.555.5096

MANAGEMENT PROFESSIONAL - RECENT M.B.A. GRADUATE
7+ Years In Sales, Facilitating Training And Handling Front Line Customer Negotiation
Top Performance In Telecommunications, Financial and Retail Industries While Completing Degree

DEMONSTRATED STRENGTHS

• Sales • Telemarketing and consulting • Team building • Motivating staffs for peak productivity •
• Managing time and resources • Problem resolution • Goal setting • Planning •

SALES AND MANAGEMENT EXPERIENCE

CONSULTANT ... BELL RINGER-PA, INC., PITTSBURGH, PA   3/97 to Present
Exceeded Sales Goals - Team Leader - Interdepartment Liaison - #1 Call Center in PA and DE

- Personally increased sales by $23,000 monthly. Exceed monthly sales goals by 15 percent as a consultant to small commercial business clients.
- Set up business productivity features, yellow page directory advertising and line installations. Complete orders by programming features into lines using SSNS, a Bell Ringer exclusive system.
- Motivate a team of 6 consultants and keep them on task via ongoing follow-up and quality reviews of consultations. Evaluate performance monthly and pinpoint areas for improvement.
- Part of a team effort that increased call center's sales by 18 percent overall - - voted #1 call center in DE and PA by an outside consulting firm.
- Liaison to focus team that improved performance and avoided overlapping of efforts - - streamlined procedures via centralizing activity reporting.
- Completed consultant training, gained direct knowledge of custoflex, centrex and pots lines. Familiar with FCC and PUC regulations governing commercial lines.
- Interface with accounting, collections and large business offices throughout the Bell Ringer region to dissect and resolve account discrepancies.

FINANCIAL SERVICES CONSULTANT ... A NATIONAL BANK, PITTSBURGH, PA   8/96 to 3/97
Front Line - Executive and General Clientele - Investment/Financial Advisor - Top Performer

- #1 in sales performance since start date. Interacted with all levels of employees to take loan applications over the phone. Opened CDs and troubleshot deposit accounts. Frequently consulted to advise best options, recommended investment plans.
- Selected to represent department to new employees during training sessions to explain benefits and services. One of 2 individuals monitoring closing documentation as a means of quality control for non-branch closings. Liaison to market-transferred employees to communicate service changeovers.

CO-MANAGER ... HOT OUTDOORWEAR, ANY MALL, PITTSBURGH, PA   3/95 to 8/96
Operations Management - Recruited, Hired and Trained Staff of 7 - Exceeded Monthly Goals

MANAGER ... ABC CLOTHING STORE, GREENERY MALL, GREENERY, PA   11/91 to 3/95
Top Sales Rep - Hiring - Training - Employee Motivation - Operations - Ad Promotions

GRADUATE ASSISTANT ... UNIVERSITY OF PENNSYLVANIA, PHILADELPHIA, PA   8/92 to 12/93
Tutor - 12 Learning Disabled Students - Special Academic Support - C.A.R.E. Program

EDUCATION

UNIVERSITY OF PENNSYLVANIA, PHILADELPHIA, PA
Master of Science in Business Administration   (Magna Cum Laude Graduate)
Bachelor of Science in Business Administration (Emphasis: Finance ... Cum Laude Graduate)

ACADEMIC HONORS, ACTIVITIES
Presidential Scholar. Dean's List. Member, Student Government. Active in intramural basketball and tennis.

54—MIS MANAGER

ZACHARY ANDERSON

15 MARSHWOOD ROAD, BRIGHTWATERS, NY 11706
VOICE ● 516-555-5555   E-MAIL ● ZACHA@AOL.COM

SENIOR-LEVEL INFORMATION SYSTEMS GROUP MANAGER

- Bottom-line oriented executive with over fifteen years of comprehensive experience in profitable oversight of teams *and* technology. Produce consistent achievements in cutting-edge information systems management, team management, contract negotiations, purchasing logistics and expense reduction. Utilize an anticipatory management style to drive results in a rapidly changing industry.

- Manage information systems group supporting production and business operations for Newsday, the nation's seventh largest newspaper. Supervise 48 direct reports and control $5 million+ operating budget. Research, assess and approve technology purchases. Implement system improvements to maintain profitable, competitive, and quality-driven production processes.

KEY MANAGEMENT AND TECHNOLOGY ABILITIES

• Advanced Technologies	• Technology Needs Assessment	• Systems Configuration
• Resource Management	• Technology Rightsizing/Upgrades	• Systems Implementation
• Project Planning/Budgets	• Win-Win Negotiations	• Parallel Systems Operation
• Project Management	• RFP Development/Review	• Intranet Development
• Expense Reduction	• Vendor Partnerships	• Technology Integration
• Team Building/Motivation	• Network Administration	• Disaster Recovery
• Presentations	• Project Lifecycle	• Systems Security
• Capital Expenditure Planning	• UNIX, SUN, LAN, WAN	• Year 2000 Solutions

SUMMARY OF MAJOR ACCOMPLISHMENTS

- Reorganized entire Newsday Business Systems operation and hardware for 20% productivity improvement and a $230 thousand savings against operating expenses. Reduced footprint, allowing department's relocation to main building, freeing old location for lucrative commercial rental.

- Recruited and led team of volunteers to independently develop Information Systems Department's Intranet Server. Program was so successful that it was adopted company-wide and became part of Information Systems Department's long-range strategic plan.

- Led project team that planned and implemented Newsday $4.8 million pagination project representing cutting-edge production technology. Project delivered ROI in only two and one-half years.

- Increased productivity and reduced expenses by purchase of new Tivoli system, replacing "manual" monitoring/installation method which provided fast-response central network monitoring, asset management and LAN central software distribution to 1,200 desktops and 35 servers.

- Developed and implemented $680 thousand network project that replaced old copper Ethernets with a fiber optic backbone and Cat5 10baseT wiring to the desktop utilizing dynamic switching hubs.

- Dramatically increased Editorial Department productivity by conversion from proprietary dumb terminals at desktops, to PC based desk-top terminals and the expansion of the Editorial front-end system from single to multiple networks.

134

54—MIS MANAGER (CONT.)

NETWORK, CLIENT SERVER, AND SOFTWARE TECHNOLOGY

- Fiber Optic Backbone
- LAN and WAN
- Wide Area Networks Through T-1 lines
- Class B Internet license
- TCP / IP
- Firewall One, Hub Watch
- Wellfleet and Cisco Routers
- Dial and Dedicated Remote Communications
- Frame Relay
- DEC 900 Dynamic Switching Hubs
- Switched and Shared Ethernet Circuits

- Tivoli LAN Management system
- UNIX Operating Systems
- SUN Solaris and Raid Disk Technology
- Sybase Relational Database
- WinFrame Syntrex for Thin Client/Fat Server
- NT Alpha Server with Microsoft Mail
- Windows 3.1, 95, NT (Client and Server)
- Microsoft Office
- Norton Anti-Virus
- Microsoft Project and Support Magic
- Total Intranet Development and HTML Code

CAREER DEVELOPMENT

NEWSDAY, INC., MELVILLE, NY   1982 to present

Director, Information Systems Group 1995 to present
Publishing Systems Manager 1984 to 1995
Editorial Project Manager 1982 to 1984

Manage 48 employees and oversee a $5.6 million budget to research, plan, install, integrate and maintain hardware, software and networks to support 2,000 internal and external customers at Newsday's New York, Long Island, and Washington DC facilities. Areas of responsibility include:

- **Publishing Group**—handles all support for print functions of publication: software, hardware, operating systems, desktop installations, upgrades, repairs 24X7 365 days.

- **Business Systems Group**—the actual operations team. Includes systems programmers and data center staff for Newsday and the Baltimore Sun. Team also handles testing and implementation of the business systems disaster recovery plan for 24 hour data center catastrophe recovery.

- **Networking Group**—the technical team. Designs, implements, installs, monitors and repairs LAN and WAN systems in New York City, Long Island and Washington DC.

Conduct strategic planning including system upgrades, and capital, operating and line item budgets. Develop RFPs, execute ROI analysis, and negotiate/review contracts. Control departmental purchasing and resource management. Track expenses and maintain budget integrity. Produce management and staff reviews, determine salary increases and payroll budget.

Lead project teams, and manage projects using Microsoft Project and Support Magic to direct teams handling internal and client hardware and software installations. Perform system reliability analysis and cost justifications on system replacements.

Head Newsday Standards Committee to determine minimum standards for desktops, servers, software packages, printers and laptops. Member of Year 2000 problem solving team. Partnered with Newsday's Electronic Publishing Department to develop an RFP and select an Internet provider. Obtained Newsday's Class B Internet license.

EDUCATION

Certificate in Information Systems Management, Hofstra University, Uniondale, NY 1994

Bachelor of Science in Business (Magna Cum Laude), Hofstra University, Uniondale, NY 1980

Lynn Charles

45 Fox Crossing ♦ Florissant, Missouri 63034 ♦ (314) 555-3333

Make-Up Artist

Motion Picture	Commercials	Print Makeup
Television	Stage Performance	Special Effects

"Lynn is an extremely knowledgeable, versatile, and gifted make-up artist. Our performers and models requested her for all their difficult make-up needs, and Lynn came through every time with professionalism and creativity. You can count on Lynn for superb artistry!"

Tom Jennings, Director
John Casablancas Modeling Center

Commercials

- **PDQ/3 Spots**
  Cast 20/12-92 Years-Old, Burdeau-Pechin Productions
  Directors/Chris Pechin & George Burdeau, North Hollywood

- **KY-98 FM Radio/3 Spots**
  Director/Al Crane - St. Louis

Films

- **CBS "CLASS ACT"**
  Director/Jon O'Brien; Producer/Brian Winter - LA

- **ABC "ABC ROCKS"**
  15 Nationwide Network Shows
  Director/David Jackson; Producer/Gary Benz - LA

Creative format and design to include graphics and script.

Resume is full of accomplishments, awards, and experience.

References are used with telephone numbers to back-up and verify the information contained on the resume.

Print & Fashion

- **Fashion Show & Showcase**
  Cast 30 Models
  Producer/Tom Kemp; Publisher/Hollywood Magazine

- **Model's Portfolios & Black/White Color Print**
  Instructor/Beauty Makeup, John Casablancas Modeling Agency

- **CPI Photo Finish; Southwestern Bell Telephone Company**

Lynn Charles   (page 2)

Print & Fashion (cont.)

- **Casting Party & Model Showcase**
  25 Models - Head Makeup Artist
  Producer/Tom Kemp; Publisher/Hollywood Magazine

- **Pure Sweat Fashion Show**
  Head Makeup Artist/Instructor
  Sponsors/Monty's Steak House, John Casablancas, Pure Sweat Clothing

- **Character & High Fashion Head Sheets**
  All Talent - Players and Others; President/Alona Perez - St. Louis

- **General Sales Catalog**
  Photographer/Rick Meoli

- **Still Photography**
  Alishe Tamburri; Photographer/Susan Ioata

- **Still Photography/Headsheets**
  Photographer/Dan Dreyfuss

Theater

- **Emmy Awards 1984**
  Makeup/Wardrobe Carol Lawrence, Fox Theater - St. Louis

Training

- **Joe Blasco Makeup Center,** Hollywood, CA
  Graduated 1984; Continuing Education 1996-Present

- **Certified Aesthetician**
  American Institute of Aesthetics

References

Joe Blasco, Joe Blasco Makeup Center	(213) 555-2222
Baron Von Mock, Enrollment Director, Blasco Studio	(213) 555-2222
Kelcey Fry, Professional Makeup Artist	(213) 555-2222
David Langford, Professional Makeup Artist	(714) 555-2222
Tom Kemp, Publisher, Hollywood Magazine	(213) 555-2222
Richard Dashut, Producer (Fleetwood Mac)	(213) 555-2222
Jack Fisher, Photographer	(314) 555-2222
Alona Perez, President, Players & Others	(314) 555-2222

Video and Portfolio Available Upon Request

56—MANAGEMENT/CEO LEVEL

C. RILEY O'BRIEN
7 Sherwood Forest Drive, Ft. Worth, Texas 76114 ● 555 555-1212

CAREER PROFILE

SENIOR EXECUTIVE experienced in the strategic planning, development and management of multi-million dollar international business operations with specific expertise in manufacturing environments. Proven record of achievement in handling newly formed joint venture corporations from start-up through profitability. Verifiable record of instituting successful "turn arounds" for troubled companies. Consistently successful in analyzing market trends and capitalizing upon market opportunities to create high-profit, high-visibility companies. Particularly adept in developing effective, loyal, and cohesive staffs. Qualifications include:

- Administration
- Communications
- Contract Negotiations
- Cost Controls
- Human Resources
- Labor Relations

- Market Analyses & Trends
- Global Perspective
- Operational Management
- Staff Development
- Plant Management
- Quality Assurance

- Operational Productivity
- Bottom-Line Profitability
- Project Administration
- Resource Planning
- Sales and Marketing
- Strategic Planning

PROFESSIONAL EXPERIENCE

E Network Technologies, Daly City, California - 1997 to present

Project Manager reporting to the Executive Vice President, Engineering and Construction with bottom-line P&L responsibility for $55 million turnkey, design, engineer, and build automatic toll collection project for California Department of Transportation (CalTrans) covering all state controlled toll bridges. Functional authority includes all aspects of human resources management, engineering, CAD, contract and subcontract administration, warranty administration, marketing plans, quality assurance, inventory control, finance, scheduling and customer relations. Functional responsibilities for systems integration include all technological aspects of RF transponders, infrared vehicle classification devices, antennae, light beams, treadles, computer controllers at each traffic lane, fiber optics, license plate capture system with character generation capability, plaza (bridge) computers including operating and report generating software, a host (central) computer system and report generating software. Ancillary responsibilities include the operation of 5 customer service centers and development of an accounting system for financial management and auditing.

- *Recruited to assume control of the most complex toll project in the western hemisphere which was over budget, understaffed, behind schedule, and undisciplined.*

- *Orchestrated complete turn around while generating $3.4 million in project cost savings in first 10 months while simultaneously negotiating schedule changes to eliminate all potential liquidated damages.*

EcoTrans Technolgy, Inc., Los Angeles, California - 1994 to 1997

General Manager reporting to the Board of Directors. Hired and trained the complete staff, using the preliminary business plan, of a joint venture between Southern California Gas Company and EcoTrans Systems, Inc., converting gasoline/diesel powered engines to compressed natural gas (CNG).

- *Perfected the business plan building the company, de novo, in a new, emerging industry.*

- *Achieved profitability within 14 months while capturing an 82% market share within Southern California.*

- *Personally generated actual sales of over $2.3 million with a 1,500+ vehicle order backlog valued at $6 million and options for an additional $3.1 million during first year.*

(Continued on next page)

138

PROFESSIONAL EXPERIENCE *(Continued)*

Yellow Bird Bus of North America, Mississauga, Ontario, Canada - 1990 to 1994

Vice President, Engineering and Sales (Consultant) reporting to the President and Board of Directors. Oversaw all phases of engineering and sales for an organization averaging $300+ million in annual sales. Responsibilities included design, development, and manufacture of heavy duty transit buses for both Canadian and American markets. Additional responsibilities included design engineering; methods and manufacturing engineering; management of the service department including technical publications, research and development of hybrid and low floor buses, and handicapped bus design; the computer aided design (CAD) group; sales, contracts, bidding, and proposals; testing and a joint development hybrid vehicle program with G.E.; and alternative fuels (CNG) design.

- *Developed a successful "turn around" business plan with other consultants. Personally developed engineering and sales and marketing plans.*

Cubic Control Corporation, San Marino, California - 1988 to 1990

International Business Development Director reporting to the General Manager. Responsible for development of both domestic and international new business. Focus was on rail and bus system new business regarding fare collection, data acquisition, and reporting systems. Handled special assignments involving overseeing Program Management, Technical Publications, and Product Support Departments for Farebox Product Line. Coordinated specifications and proposal development with consultants worldwide.

- *Analyzed new market segment (light rail), developed and prepared the proposal strategy and direction to win new job in Hong Kong.*

- *Established a nationwide team of efficient, effective sales representatives.*

- *Captured 80% of Canadian electronic fare box market share.*

Hagen Corporation, Beaumont, Texas - 1986 to 1988

President and Chief Executive Officer. Set-up the Sales and Marketing and Engineering Departments for this newly acquired company building patented shuttle buses with optional wheelchair lifts. Established the computerized financial and inventory control systems. Coordinated efforts of the Sales and Marketing, Service, Spare Parts, and Engineering Departments in Western States. Assumed responsibility for operational management of 4 departments of a $375 million corporation including: Technical Publications, Contract Administration, Marketing Analysis, Marketing Administration, Public Relations, and Advertising. Prepared near-, intermediate-, and long-range marketing plans.

- *Converted company from loss to profitability within first 14 months.*

- *Increased market share from 12% to 30% while competing with six competitors.*

- *Directly responsible for developing bids and proposals, leading to negotiating 81 contracts and establishing $175 million business backlog of buses to be built to U.S. Government (ADB) specifications.*

EDUCATION

Southwestern Texas Institute of Technology, Alice, Texas
Bachelor of Science - Industrial Management and International Business

57—MANAGEMENT CONSULTANT/E-MAIL RESUME

Renee Christian
2724 Cameron Street
Washington, DC 20008
W (703) 555-1212, H (202) 555-1212
renee.christian@mail.argi.net

Simple layout and font designed for e-mail and transport via the Internet.
Simple font that is monospaced.
No elaborate formatting, bold, underlined, or italicized fonts.

EXPERIENCE

A.T. Kearney - EDS Management Consulting Services
Washington, DC
Principal, 1996 - present
Senior Manager, 1994-1996

Specialized in the development of business plans for emerging
wireline and wireless telecommunications companies.
Responsible for client relationship management and sales into
pre-existing client base. Responsible for managing large (up
to 30 people) work teams. Provided advice to senior
management on a broad variety of strategic and operational
subjects, including:
-Business strategy and planning
-Financial modeling
-Competitive assessment
-Market segmentation
-Operations design and optimization
-Mergers and acquisitions
-Implementation assistance
-Churn management

Deloitte & Touche Management Consulting
Washington, DC
Manager, 1993-1994
Senior Consultant, 1991-1993

Primarily focused on developing market entry strategies for
telecommunications clients. Developed very strong financial
modeling skills. Responsible for managing smaller projects
and developing client deliverables.

EDUCATION

B.S. in Physics, Montana State University, 1989
MBA in Finance, Carnegie Mellon University, 1991

PUBLICATIONS / SPEECHES

"An Ice Age is Coming to the Wireless World: A Perspective on
the Future of Mobile Telephony in the United States," 1995
Thought Leadership Series, EDS Management Consulting

"Analysis of the FCC's Order Regarding Ameritech's Application
to Provide InterLATA Toll Services Within the State of
Michigan," A.T. Kearney White Paper, 1997

Presented at over two dozen industry conferences, workshops, and
panels

140

58—MANUFACTURING MANAGER

CHERYL R. MICHELLES
151 W. Passaic Street, Rochelle Park, New Jersey 07662
(201) 555-3772   Email: xxxxxx@prodigy.com

MANUFACTURING MANAGER

production planning   purchasing   inventory control

PROFILE SUMMARY & STRENGTHS:

A skilled professional with a strong desire to contribute to the productivity and proficiency of a progressive organization. A conscientious individual with a strong work ethic and the ability to interface with others at all levels. Detail oriented; possesses strong analytical skills used in effective decision-making. Able to work independently and within fast paced environments.

COMPUTER SKILLS:

Internet   Windows 95   AMAPS   MPS II   Lotus 1-2-3   Lotus cc:Mail
Ami Pro WordPerfect Freelance Graphics AS/400 JD Edwards RMS System

CAREER HIGHLIGHTS:

At Reckitt & Colman achieved a 1200% reduction in line order cuts, exceeded objectives for inventory turns, and reduced total inventory by 15%.
   Spearheaded production planning, component ordering and coordination of start-up with co-packers for 7 new products.
   Minimized scrap losses through disposition of on-hand inventory during formulation and packaging changes, and sales of inventories to contractors.

At Castrol North America Automotive identified and recovered $1.2 million in supplier overcharges through the development of computer models.
   Saved over $500,000 through price renegotiations with vendors.
   Devised and maintained cost file spreadsheets by utilizing Bills of Material and costs to determine overall case cost of each product.
   Served as team member of Cross-Training Program, Supply Chain Management Team and numerous task forces.

PROFESSIONAL EXPERIENCE:

RECKITT & COLMAN, Wayne, New Jersey   1996 - Present
Materials Administrator - External Manufacturing

Recruited to a high profile position accountable for master scheduling and production planning for 11 co-packers (subcontractors) at 14 locations nationwide for a major consumer products company.
Oversee materials ordering and raw material inventory management at 9 co-packers.
Manage $1.5 million monthly projected inventory balances for finished goods inventory; ensure attainment of corporate inventory goals and objectives.
Accountable for 80+ sku's and 20+ different product types including powders and liquids.
Actively participate in new project management.
Utilize AMAPS and MPS II computer systems in daily operations.

CHERYL R. MICHELLES   Page 2   (201) 555-3772

PROFESSIONAL EXPERIENCE continued...

CASTROL NORTH AMERICA AUTOMOTIVE, Wayne, New Jersey   1990 - 1996
Supply Coordinator, Contract Manufacturing (1993-1996)
Assistant Supply Coordinator, Packaging (1992-1993)
Associate, Marketing Department/Supply Department (1990-1992)

Achieved fast track promotions in recognition of superior performance.
Accountable for the development of contract manufacturing supplier strategies for a company specializing in automotive aftermarket products; managed vendor sourcing, price negotiations, packaging development and internal coordination.
Supervised material valued at $80 million; held purchase order signature authority to $20,000 and invoice approval authority to $100,000.
Procured $6.0 million annually in products and services.
Monitored contractor/supplier performance against strategy plans and implemented corrective actions.
Controlled raw materials and finished goods inventory at sub-contractor locations.
Provided logistical guidance for 3 plant coordinators and material requirements at 7 contract packers.
Accountable for Canadian production requirements, purchase order processing and sales order entry following Canadian office integration into US operations.
Oversaw installation of JD Edwards software and trained suppliers in application use.

PETRIE STORES CORP., Wayne, New Jersey   1987 - 1990
Assistant Store Manager, Marianne's (1988-1990)
Acting Store Manager, Stuart's (1987-1988)

Oversaw store's daily operations; accountable for sales promotions, customer service, payroll, floor plan changes, scheduling and staff development.

EDUCATION:

Bachelor of Science in Marketing
William Paterson University, Wayne, New Jersey
Cum Laude

Continuing Education in:
• Materials Management Certification program - Bloomfield College
• APICS Inventory Management and Control
• GMP/FDA Training
• Karass Effective Negotiating
• Zenger-Miller Core Interpersonal Skills

Regularly attends APICS conferences

PROFESSIONAL AFFILIATIONS:

American Production and Inventory Control Society (APICS)

59—MARKETING DIRECTOR

> *Summary at the front end does a great job of quickly summarizing the key elements of the marketing professional's experience.*
>
> *The bullets in the Management Experience section focus on accomplishments, quantifiable when possible.*

MARKETING DIRECTOR

Jeff Kilpatrick

2 West Park City Drive
Dallas, TX 75222

(214) 555-2597

Professional Profile

Management professional attuned to the ever-changing needs of business. Extremely service-oriented with a unique combination of intuitive and analytical abilities. Astute in identifying market plan needs, creating actionable programs and effectively interacting with the sales field.

- Segmentation Targeting
- Financial Forecasting
- Retail Sales
- Collateral, Displays and
- Total Communications Strategist

- Branding
- Advertising
- Sales Promotion
- New Product Rollout
- Strategic Planning

Management Experience

Proctor and Gamble, Cincinnati, Ohio
Marketing Director (paper products division)
Fortune 100 consumer products company

Positions held: Marketing Director (1992 to present); Manager Market Development (1988 to 1992)

Selected Achievements:

- Conducted research and initiated new product line in paper products, including product specification development, segmentation identification and rollout plan. New line gained a 32% share within six months, nearly the P&G average for toiletry products.

- Identified and developed new distribution channel for P&G paper products line, which included retail recruitment, rollout schedule, inventory management and training schedule. New channel represented a 7% share of distribution mix within 2 years.

- Recommended shift in brand image to target younger buyers, which was successfully implemented and improved name awareness with that segment by 12%. Overall brand awareness for individual product lines reached 93% in 1997.

- Dedicated extensive time and energy to improving communications with personnel, accepting full responsibility for scheduling, performance reviews, employee motivation, and boosting morale. ESS feedback for our group was in the top 20% in all of P&G.

Additional Experience
Advertising (4 years) for Chiat Day, specializing in the automotive and hospitality industries. Specifically, worked as account manager for clients including AirTouch, Volvo and Pepsi-Cola.

Education
MBA, University of Texas-Austin   Graduate 1985
Bachelor of Arts, Marketing, Vanderbilt University, Nashville, Tennessee, 1980

60—MEDICAL RECORDS CLERK

MELISSA NYDIR

1154 ALEXANDER COURT, SOMEWHERE, MI 19123   **123.456.7890**

MEDICAL RECORDS CLERK

Self Starter With 11 Years Experience In Accessing Patient Records For Hospital And Private Practice
Improved Operations, Earned Quality Improvement Award, Maintained Strict Confidentiality

HIGHLIGHTS

- 8+ years as a front line liaison to patients, physicians and attorneys to authorize and release medical records for a major metropolitan medical center.
- Special knowledge of medical terminology and surgical procedures, anatomy and physiology, as well as primary and third party insurance regulations.
- Computer literate. Experienced with Word, WordPerfect for Windows and DOS, Lotus 1-2-3, rBase and offer strong understanding of internet applications.

PROFESSIONAL EXPERIENCE

MEDICAL RECORDS CLERK ... METRO MEDICAL CENTRE, DETROIT, MI   **2/90 to Present**
Train New Hires - Earned Top Evaluations - Authorize Release Of And File Patient Records

- Handle chart retrieval for up to 200 inquiries daily for a 400-bed, 12-department regional suburban hospital and oncology clinic.
- Gained a reputation for accuracy and timely responses to requests - - most frequently requested clerk by physicians' and attorneys' offices to locate and process charts.
- Regulate access by insurance carriers - - ensure proper channels have been followed to maintain the strictest of confidentiality.
- Secure patient authorizations for release of records and document activity and clearances using a Windows WordPerfect-based application.
- Interview new mothers to discuss payment and to verify services provided. Forward questionable claims for immediate correction to billing department.
- Asked by director to train new hires - - oriented 6 clerical personnel to date, #1 person of a staff of 11 consulted for advice and answers by my peers.

ACCOMPLISHMENTS:

- On numerous occasions, pinpointed flaws in completion of records, which resulted in better accuracy in reporting and improved quality overall.
- Earned a Quality Improvement Award for identifying a discrepancy in reporting practices that facilitated JCAHO Certification and avoided $45K in fines.
- Received a thank you from a general practitioner for resolving a billing discrepancy due to improper categorization discovered while performing deficiency analysis.

MEDICAL SECRETARY ... JOSEPH E. ELDER, M.D., THORNBURG, MI   **4/87 to 2/90**
Medical Records - Filing - Medicare and BC/BS Claim Submission - 5-Location General Practice

- Restructured charting setup and redefined order of documentation, reducing time spent by billing staff in reviewing and approving benefits for a satellite office that was implemented throughout a 5-office network.
- Slashed excess time in locating referrals by documenting their location and whether they were received via fax or mail from other providers. Previously, referrals were not organized in any fashion.
- Took the initiative to anticipate mailings to regular referring network of physicians - - preprocessed shipping information and decreased preparation time.

EDUCATION

YPSLANTI BUSINESS SCHOOL, YPSLANTI, MI
Medical Terminology Certificate

61—MENTAL HEALTH PROFESSIONAL

TAKKUHA PATEL, C.S.W.
151 W. Passaic Street
Rochelle Park, New Jersey 07662
(201) 555-3772

MENTAL HEALTH PROFESSIONAL

An experienced and knowledgeable social worker with demonstrated skills in management and administration. Excels at fostering communications and ensuring smooth operations. Served as a Resident Companion Volunteer (Geriatrics) at Bergen Pines County Hospital. Experienced in implementing community programs to increase awareness and promote education.

PROFESSIONAL EXPERIENCE

ASSOCIATION for RETARDED CITIZENS, Hackensack, NJ (1994 - Present)

Counselor

Train and supervise 5 - 7 clients in daily living skills.
Maintain detailed logs on a daily basis for each individual.
Administer medications according to physicians' orders.
Maintain continuous flow of communication to determine clients' status in Day Program.
Interface with State representatives to ensure adherence to laws and regulations.
Communicate and schedule client sessions with healthcare professionals; accountable for overseeing medication changes according to physicians' orders.
Confer with inter-office personnel and senior management; instrumental in strategy development and implementation.
Update family members/legal guardians on clients' status and development.
Formulate and develop Individual Rehabilitation Programs (IHP); devise and implement clients' daily living skills and recreation programs.
Contribute to the development of behavior modification plans for clients.

SELECTED HIGHLIGHTS

Instrumental in the development of community treatment satellites.
Participated in researching and authoring 2 grants totalling $240,000, including a comprehensive prevention program.
Team member actively involved in the formulation of Quality Assurance procedures and policies.

EDUCATIONAL BACKGROUND

New Jersey State Certified Social Worker (C.S.W.)

B.S.W. in Psychology/Sociology, Government Arts and Science College, Gulbarga, India

Additional training:
Developmental Disabilities   Prevention of Abuse and Neglect   Leadership Development
First Aid   CPR

62—MERCHANDISING MANAGER

> "BIG" heading clearly indicating her focused objective.
>
> Convincing credentials supported by stalwart achievements.
>
> Education is creatively centered in the page to balance strong Profile and Experience sections.

LINDA J. WESTBROOKE

• Uniquely Qualified For Merchandising Manager Positions •

115 North Union Boulevard, Colorado Springs, CO 80909 • 719/ 555-9050 • ideaboss@aol.com

CAREER PROFILE

- Top-performing senior executive with unique tactical and strategic planning skills
- Strong ability to identify consumer trends and create new sales opportunities
- Proven record of leading organizations through growth and expansion periods
- Highly skilled at managing licensed properties
- Demonstrated competence in managing departmental budgets
- Able to deliver significant revenue and profit gains within competitive markets
- History of igniting stagnant and declining product lines into profitability
- Experienced in working within the constraints of limited resources to effect change

FORMAL EDUCATION

Post Graduate studies, Mass Marketing, University of Chicago
Bachelor of Arts, Fashion Design, University of North Texas, Denton, Texas
Associate of Arts, Design, Paris Junior College, Paris, Texas
Graduate, Managing Innovation, Stanford University
Graduate, Executive Edge Management, University of Chicago

RELEVANT PROFESSIONAL EXPERIENCE

Vice President - New Product Design - Merchandising Manager   *(1984 to Present)*
(A Fortune 500 manufacturer of children's toys and dolls)

Manage product design and development across various categories. Direct forecasting and administration of a $5 million annual operating budget. Responsible for long-range strategic planning and product development. Direct all worldwide product merchandising activities. Oversee the worldwide new product acquisition and development process. Interface with inventors and vendors in an on-going search for innovation. Supervise quality control from ideation through manufacturing. Serve as company spokesperson both nationally and internationally. Direct media relations activities.

◆ **Special Achievements and Awards:**

- ◆ Increased sales from $400 million in 1987 to $1.1 billion in 1994
- ◆ Created numerous new product line extension strategies
- ◆ Instrumental in obtaining a 5-year exclusive contract with Disney
- ◆ Direct the design team's winning pitch for all brand name merchandise
- ◆ Grew one product line in 3 months from $26 million to over $40 million in sales
- ◆ Championed the use of internal concepts with outside concepts
- ◆ 1996 Chairman's Award
- ◆ 1995 President's Award

REFERENCES AND FURTHER DATA UPON REQUEST

63—MILITARY CONVERSION

J. C. Smith

7 Del Mar Drive, Oceanside, California 92054 ☎ 555 555-1212

"...a 'one-of-a-kind' officer. The absolute **best** Major on this Base. He is the best I have ever served with in any capacity. Top 1% of all Majors I know and #1 of 30 Majors at this Base."
Lieutenant General I. M. DeBoss USMC
November 10, 1995

CAREER PROFILE

Confident, dependable, versatile management professional with extensive diverse experiences enhanced by graduate level studies. Global perspective based upon assignments and travel abroad. Articulate, sagacious, problem solver with superior analytical and communication skills. Organized, meticulous, and methodical: particularly adept at problem identification, research, analysis, and resolution. Qualified by:

- An established record of progressively responsible positions of trust at the highest levels of government,
- A proven and verifiable history of success in the administrative management of military units, and
- An innate ability to develop loyal and cohesive staffs dedicated to the task at hand.

COMPETENCIES

- Communications Skills
- Computer Systems
- Human Resources
- Leadership & Supervision
- Operations & Planning
- Organizational Skills
- Project Management
- Staff Coordination
- Time Management

EDUCATION AND PROFESSIONAL TRAINING

University of Phoenix (Satellite Campus), Vista, California
Master of Arts in Organizational Management

Southern Illinois University, Carbondale, Illinois
Bachelor of Science in Aviation Management
Summa Cum Laude

PROFESSIONAL EXPERIENCE

United States Marine Corps - 1972 to present

Major - Administrative Officer whose professional human resources management responsibilities expanded to include areas of consolidated finance and accounting, career development, computer information systems, international diplomacy and protocol, senior staff coordination, staff training and development, personnel services, equal opportunity, and security.

Significant accomplishments:

- Awarded prestigious Meritorious Service Medals for superb performance of duty, 1986 and 1992.
- Reduced the MCDOSET error rate from 43% to 1.5% within 36 months.
- Personally managed a $1+ million annual budget of error-free procedures, 100% accountability, and faultless internal procedures.

AFFILIATIONS AND MEMBERSHIPS

Marine Executive Association, Washington, DC
The Retired Officers Association, Washington, DC

Appropriate personal and professional references are available.

Internet- and scannable-friendly resume—no lines, accentuations, or graphics. Military conversion resume that successfully highlights high-tech skills used in the military that have a direct application and value in civilian work.

Nathan G. Hale
11749 North O'Connor Road
Irving, Texas 75048
972-555-7697

UNIX Systems Administrator

System Availability / Resource Security / Workload Control / User Training

Innovative professional with impressive record of achievement in technical and systems support. Develop procedures to streamline daily operations. Comfortable communicating at all levels from the Pentagon to coworkers and trainees.

Practiced Expertise:

Access control • Resource authorizations and permissions • System maintenance and updates
Peripheral hardware installation and configuration • performance monitoring • user account audit
system backup • recovery and user support

Technologies:

AT&T 3B2/600G • GTSI 433DX/B • KG-84 Crypto Equipment • AT&T UNIX System V
CAS-B CAD • Windows 95 MS Office 95 Pro • Harvard Graphics • Power Point

Professional Experience:

United States Air Force - Lackland Air Force Base Annex   San Antonio, Texas

UNIX Computer Systems Administrator   1995 to Present

Control all operational programs on AT&T UNIX System V.

- Maintain mission-critical computer resource in support of Air Force Operations.
- Improved workflow, reduced time-on-task and enhanced system security by implementing customized system's user interface.
- Identify potential system failures by running built-in diagnostic routines.
- Coordinate with Field Engineers to correct equipment malfunctions.
- Schedule and perform preventive maintenance.
- Established on-the-job training program to educate users in efficient system use and mission-critical software programs.
- Control the movement of munitions and insure assets are properly tracked, maintained and delivered.

Specialized Training:

CAS-B 3R2 System - 82nd Training Group - Sheppard AFB, Texas - May, 1997
Munitions Systems Specialist - Lowry Training Center - Lowry AFB, Colorado - September, 1994

Awards:

Armed Forces Expeditionary Medal awarded for outstanding service in support of Operation Southern Watch, 4404th Tactical Fighter Wing (Provisional), Dhahran, Saudi Arabia - 1996.

65—NUCLEAR SCIENTIST

NUCLEAR SCIENTIST / ENGINEER
Research / Engineering Expert
Ph.D. MIT

Hans Von Pels
1834 South Wellington Circle
Portland, OR 55555
(555) 555-7892

BACKGROUND SUMMARY

Offering 18 years of comprehensive research, analysis, theory development, and practical application in the area of Radiation Protection for the United States Government through a long-term grant provided by the University of Oregon. Evaluate/monitor activities at six nuclear power plants, eleven weapons test sites, and six research centers - all dealing with nuclear/radiation technologies.

A global visionary recognized for "out-of-the-box" thinking in anticipating problems before they arise, and troubleshooting/problem resolution when challenges do occur. Link solid background in scientific theory with today's practical applications in a safe, environmentally sound manner consistent with long term consequential projections.

HIGHLIGHTS OF PROFESSIONAL EXPERIENCE

THE COLDWELL-BECKER PROJECT - US GOVERNMENT FUNDED PROJECT
(Through a Special 30 Year Grant at the University of Oregon)
Portland, OR / 1980 - Present

NUCLEAR SCIENTIST / RESEARCH ENGINEER
Monitor nuclear/radiation levels for nuclear power plant facilities, radiation testing/research centers, and underground test sites.   Monitor/predict radiation levels to assure public/environmental safety. Evaluate radiation fields, instrument selection and performance - investigating erratic/unstable radio-analysis and external/internal dosimetry results in relation to accepted levels of deviation. Measure plant/test center emissions and environmental risks and propose appropriate solutions. Scrutinize all facilities for regulatory compliance and adherence to stringent safety specifications and blueprints.

Work closely with government regulators, national and international environmental organizations, management, university oversight committee, facility managers and executives, and other scientists in achieving measurable results within extreme minimal risk conditions. Submit key, comprehensive monthly activity reports with accompanying recommendations. Present oral analysis to Congressional Subcommittee in Washington on an annual basis on behalf of The Coldwell-Becker Project. Train new scientists to the project and work closely with university interns assigned to the project.

- Successfully developed a Radio-incremeter Model to predict future radiation based on 40 years of scientific data. Model had helped decrease radiation leakage nearly 98.7% at testing and research centers.
- Upgraded and enhanced practical beta radiation protection device that has been in use for the past 12 years without incident. Device has saved more than $102 million in corrective action as a result of accidental malfunction liability.

Hans Von Pels
Page two

HIGHLIGHTS OF PROFESSIONAL EXPERIENCE (Continued)

- Designed and implemented state-of-the-art mathematical model using PredictorTask software to predict future malfunctions at nuclear power plants
- Introduced Bio-voluminous feedback for emergency response to potential hazards and prepared policy manual for disaster control/emergency response to potential mishaps - currently being used by all 50 states.
- Authored 24 in-depth articles for major publications over past ten years including:
    - *Radiation Management and Control,* Scientific Digest, 7/98
    - *The New Age of Power Source Technology,* Smithsonian Report, 3/98
    - *Models of Prediction for radiation Disaster Management,* US News and World Report, 9/97
    - *Environmental Threats From Third World Nuclear Energy Sources,* Time Magazine, 4/97

EDUCATION

MIT, Cambridge, MA
Ph.D. in Nuclear Physics

HARVARD UNIVERSITY, Cambridge, MA
Master Degree: Nuclear Physics

TUFTS UNIVERSITY, Boston, MA
Bachelor of Science: Physics

* Possess sophisticated computer/programming skills
* Fluent in English, German, French, and Spanish
* Consultant for US Government to Poland, Czechoslovakia, and Romania

- References and Portfolio Furnished Upon Request -

66—NURSE

JACQUELYN A. LOGAN. RN, CNOR
7402 Field Park Drive South
Athens, OH 45701

(614) 555-9019

CERTIFIED REGISTERED NURSE - OPERATING ROOM
Training, Supervision, and Leadership Skills

A highly skilled Nursing Professional with qualifications in management and expertise in OR nursing. A hospital advocate, promoting the hospital by providing high quality nursing care and unparalleled patient service. A cost-containment professional through quality management and inventory control. Provide total back-up and support to physicians and surgeons.

STRENGTHS & SKILLS

OR Specialty; Vascular, General, Thoracic, Renal
Scrub Experience
Experienced Circulator
Organization and Time Management
Adhere to High Ethical Standards; Professional Integrity

Personnel Training and Supervision
Assessment Skills
CPR Certified
Educator to Patient, Family, Peer Group
Communicating Skills - Verbal / Written

PROFESSIONAL EXPERIENCE

KENT COUNTY MEMORIAL HOSPITAL, Shade, OH — 1990 - 1997
Circulating Nurse Coordinator (Staff OR Nurse rotating to all specialties)
♦ Developed organizational system for effective management of cases, personnel, and surgeon start-time
♦ Reduced operating suite turnover time from 35 minutes down to 8 minutes
♦ Participant on Cost Containment Committee; reduced expenses without compromising quality of care

SOUTH COUNTY / ROGER WILLIAMS HOSPITALS, Athens, OH — 1988 - 1989
OR Nurse - Per Diem (While attending school at Ohio University)

THE PLAINS OHIO UROLOGICAL ASSOCIATES, The Plians, OH — 1986 - 1988
Registered Nurse (Office / OR)
♦ OR Coordinator; pre /post-op educator; post-op visitations from lithotripsies (ESWL)

ROGER WILLIAMS GENERAL HOSPITAL, Athens, OH — 1978 - 1986
Registered Nurse / Specialty Nurse
♦ RN in pilot project; primary patient care, providing continuity of pre/post op care; charge experience
♦ Specialty nurse responsible for orientation and staff development; coordinated daily schedule of OR

WOMEN & INFANTS HOSPITAL, Athens, OH, RI — 1974 - 1978
Registered Nurse / Managerial Position - 30 beds / 30 newborns
♦ Hired for pilot project: Family centered care; expanded project to entire post partum unit
♦ Educated parents / staff with hands-on demonstration care of newborns; nursing and nutritional guidelines

EDUCATION & TRAINING

Newport Hospital School of Nursing, Newport, RI — **Diploma, 1974**
University of Rhode Island, Providence, RI — **BSN/Masters Program, Currently Enrolled**

AFFILIATIONS

Member / Board Member: AORN
Delegate to National Congress - AORN, 1996
Member / Supporter of MADD

Member: National Kidney Association
Century Club Contributor: AORN Foundation
Active in HCFA - Campaign to oppose rule changes

67—NURSE ANESTHETIST

Melanie Johnston, CRNA

1555 Main Street • Charleston, WV 25302 • (304) 555-8443

"... exhibits high degree of intelligence and readily grasps new concepts ... has an affable charm ... interacts well with patients, colleagues ... even in the most stressful of situations."

James McCroskey, MD
General Anesthesia Services
Charleston, WV

"... reliable and responsible team player ... willingly shares the workload ... level-headed and competent in an emergency ... proficient and knowledgeable in anesthetic skills and techniques."

Barb Schmitt, CRNA
CAMC-Memorial Division
Charleston, WV

"... an excellent anesthetist who remains calm under pressure ... highest integrity ... exhibits excellent leadership ... has been a tremendous asset to our organization."

Lee Ann Smith, CRNA, BA
Instructor, CAMC School of
Nurse Anesthesia
Charleston, WV

"... a very responsible employee ... always volunteering for additional assignments ... prompt and punctual ... has a positive attitude ... a valuable asset to our staff."

Tamy L. Smith, Charge CRNA
CAMC-Memorial Division
Charleston, WV

Professional Profile

- **Certified Registered Nurse Anesthetist**
- Bachelor's degree and four years CRNA experience
- Clinical instructor with over 1,000 hours experience
- Outstanding clinical expertise and proficiency
- Attend weekly continuing education meetings
- Excellent problem solver who works well under pressure
- Reputation as a team player with superb people skills
- Upbeat, personable, and highly energetic

Licensure & Professional Affiliations

- Certified Registered Nurse Anesthetist, Certificate #22250
- Registered Professional Nurse, License #0556500
- Member, American Association of Nurse Anesthetists
- Member, West Virginia Association of Nurse Anesthetists
- Professionally involved with local Women's Health Center

Professional Experience

CHARLESTON AREA MEDICAL CENTER MEMORIAL DIVISION
Charleston, West Virginia   1984-Present
- □ **CRNA - Cardiovascular Center** - 1993-Present
  Surgeries include arterial bypasses, hearts, amputations, gall bladders, mastectomies, biopsies, major orthopedics
- □ **RN - Medical Surgical** - 1984-1993
  RN and charge nurse duties on a 40-bed med/surg unit, included adolescent ward and peritoneal dialysis

Education

- **Certificate of Anesthesia,** Charleston Area Medical Center School of Anesthesia, Charleston, WV, 1989
  □ Received Josephine A. Reier Memorial Scholarship Award
- **Associate Degree in Nursing,** University of Charleston School of Health Sciences, Charleston, WV, 1983
  □ Received Nursing Student Achievement Award
- **Bachelor of Fine Arts Degree,** *magna cum laude,* Arizona State University, Tempe, Arizona, 1977

68—NUTRITIONIST

CYNTHIA L. CREEK

115 North Union Boulevard, Colorado Springs, Colorado 80909

719/555-9050

NUTRITIONIST

Dietary Specialist / Lite Fair Specialist / Presentation Winner

SUMMARY OF QUALIFICATIONS

- Over 10 years of progressive and diversified experience in the food service industry
- Comprehensive experience in all aspects of menu design and banquet planning
- Proven professional in planning, budgeting, coordinating and scheduling
- Experienced in fast-paced and high-volume environments
- Broad experience in communicating and interfacing with vendors
- Unique record of flexibility and adaptability to any assignment or position

FORMAL EDUCATION and LICENSES

Bachelor of Science, Biology, University of Nebraska
Licensed Nutritionist and Dietitian, State of Colorado

PROFESSIONAL EXPERIENCE

December 1995
to
Present

NUTRITION SERVICES SUPERVISOR
PENROSE ST. FRANCIS HOSPITAL, COLORADO SPRINGS, COLORADO
- Manage a public access cafeteria serving 1800 meals daily
- Direct patient care nutrition services for a 620-bed hospital
- Supervise, schedule, train, motivate and evaluate a 23± person support staff
- Responsible for food and labor cost control budgets and variance analysis
- Coordinate and interact with nursing staff regarding special patient dietary needs
- Update and maintain accurate patient records utilizing a sophisticated database
Specific accomplishments:
   Reduced departmental food and labor costs by 22%
   Reduced turnover by 400%
   Honored for 'Lite Fair' culinary skills (Nutritional Services Int'l.)

December 1993
to
December 1995

PANTRY CHEF
GARDEN OF THE GODS COUNTRY CLUB, COLORADO SPRINGS, COLORADO
- Prepared all side, entree salads, catering trays and desserts for this 5 Star facility
- Served as Special Assistant to the Chef Garde Manager
- Organized and coordinated the setup and presentation of banquet tables
Specific accomplishment:
   Blue Ribbon Award for Presentation Skills (Taste-of-the-Rockies)

May 1990
to
November 1993

KITCHEN MANAGER
INLAND BOAT CLUB, SACRAMENTO, CALIFORNIA
- Planned and prepared California style cuisine at this fast-paced resort
- Developed shift schedules, trained all employees and prepared all menus

REFERENCES AND FURTHER DATA UPON REQUEST

69—OPERATIONS MANAGER

WILLIAM B. FRANKLIN

223 Dahlia Road
Garden Grove, New York 10576
(516) 555-0988

SENIOR OPERATIONS MANAGER
Cross-Functional General Management & Operations Management Expertise

---

Combines 10+ years of progressively responsible business experience in:

- Strategic Business Planning
- Information & Telecommunications Systems
- Multi-Site Facilities Management
- Capital Acquisition & Purchasing

- Vendor Selection & Negotiations
- Cost Reduction & Profit Improvement
- Quality & Performance Improvement
- Project Planning & Management

PROFESSIONAL EXPERIENCE:

INTERNATIONAL BEVERAGE CORPORATION, New York, New York   1985 to Present

Fast-track promotion through increasingly responsible operations management positions with the fourth largest manufacturer/distributor of specialty beverage products in the U.S. Advanced based on consistent success in process improvement, cost reduction/control, multi-site operations management and interdepartmental relations. Career highlights:

General Manager / Office Manager (1989 to Present)

Senior Operations Manager directing all general management functions for corporate headquarters. Scope of responsibility is diverse and includes manpower planning/scheduling, vendor sourcing/negotiations, supply and capital equipment purchasing, and interdepartmental relations. Liaison between Chairman and executive management team. Additional responsibilities and achievements:

Real Estate & Facilities Management

Control $39.5 million in company-owned real estate assets (e.g., manufacturing and warehousing facilities, office space, retail operations). Manage the complete project cycle from initial design and estimating through planning, scheduling and site supervision. Hold additional responsibility for materials planning, purchasing, vendor contract negotiations and inventory control. Coordinate project scheduling to minimize impact on daily business operations.

- Facilitated the construction/capital improvement of 28,000 sq.ft. multi-use commercial office building. Negotiated leases with outside tenants and achieved 90% occupancy in nine months.

Telecommunications & MIS Affairs

Oversee the planning and implementation of all MIS and telecommunication operations for corporate headquarters, manufacturing plant and 12 company-owned retail stores.

- Facilitated conversion to state-of-the-art telecommunications system including 100+ phone lines, voice mail, credit card processing and emergency/security systems. Identified department needs, prepared specification package, selected vendor and facilitated implementation.
- Renegotiated local and long distance carrier services for $12,000 savings in annual telecommunications costs.

154

WILLIAM B. FRANKLIN - *Page Two*

Capital Equipment & Leases

Manage $1.7 million in capital equipment and technology assets. Identified and sourced vendors, and negotiated multi-year contract agreements for $36,000 savings to the corporation annually. Hold full responsibility for the acquisition, maintenance, security and repair of corporate fleet.

- Renegotiated leasing contracts and saved 12% in annual leasing costs.

Human Resource Affairs

Work in cooperation with department heads to evaluate staffing needs in response to increased productivity and performance demands. Active participant in recruitment, selection, training and evaluation of employee performance at the corporate office.

- Guided development of internal staffing and manpower planning programs to meet headquarters, sales and operating needs.

Finance & Administration

Oversee the design and implementation of budgeting, general accounting, financial analysis and financial reporting systems to meet the changing requirements of the corporation.

- Expanded communications and information exchanged between Chairman, executive management team and department managers to enhance budgeting and internal controls.

Warehousing & Production Supervisor (1985 to 1989)
Quality Assurance Technician (1985)

Recruited to International Beverage Corporation to direct the entire product quality cycle for four simultaneous processing operations. Promoted within five months to Warehouse Supervisor and within six months to Warehousing & Production Supervisor. Held full responsibility for the planning and management of all plant operations (e.g., production control, scheduling, quality, testing, shipping/receiving, materials, inventory, staffing, financial reporting) for a 48,000 sq.ft. manufacturing and warehousing facility.

Introduced a series of productivity improvement, process reengineering, cost reduction and performance management programs that consistently improved production output, product quality and customer satisfaction. Innovated unique solutions to complex operating problems.

- Introduced raw materials recycling program which reduced scrap by $50,000+ annually.
- Increased warehouse capacity by 10% through improved storage techniques and effective utilization of space.
- Introduced new systems and processes to reduce reliance on manual labor for transfer of raw materials from truck to warehouse. Increased productivity by up to 25%.
- Appointed Safety Director in 1988. Reduced lost time accidents by 10% within two months.

EDUCATION:

B.A., Business Administration & Labor Relations, 1991
CITY COLLEGE OF NEW YORK

Completed 62 hours of continuing professional education in Operations Management, Telecommunications, Real Estate and Construction Management.

70—PARALEGAL

Kathy Glass

143 Hillside Lane • Belleville, Illinois 62223 • (618) 555-7770

Paralegal/Legal Administration
Legal Investigations...Personal Injury...Family and Criminal Law

Solid training and experience in legal administration and paralegal studies involving research, writing, investigation, litigation support, mediation/arbitration techniques, rules of evidence, and dissolution of marriage proceedings. Familiar with word processing, spreadsheet, docket control, and database applications. Key responsibilities included:

- Drafted and filed motions, interrogatories, initial and responsive pleadings.
- Tracked client employment records and other information using various sources.
- Helped clients fill out income, expenses, and property statements.
- Assisted with research for preparation of briefs.
- Prepared settlement agreements and other closing documents.

Education

ST. LOUIS COMMUNITY COLLEGE AT MERAMEC
Certificate in Paralegal Studies, DECEMBER 1998

BELLEVILLE AREA COLLEGE, BELLEVILLE, IL
Associate of Science Degree, Business, DECEMBER 1994
Associate of Arts Degree, Psychology, MAY 1993

Legal Experience

U.S. ATTORNEY'S OFFICE, SOUTHERN DISTRICT OF ILLINOIS, SEPTEMBER -DECEMBER 1998
- Completed a 100-hour internship under the supervision of a paralegal supporting 40-50 attorneys in both criminal and civil cases.
- Observed and studied legal proceedings associated with the recent Venezia trial. Took detailed notes; organized exhibit list.
- Summarized depositions, videotapes, and medical records.
- Reviewed docket cards and prepared trial notebooks.

LORRI MOTT, ATTORNEY, JULY 1997-SEPTEMBER 1998
- Worked extensively with clients in person and via telephone.
- Drafted pleadings, affidavits and prepared discovery requests.
- Initiated and organized files for trial.
- Typed correspondence and billing information.

ADDITIONAL EXPERIENCE: Waitress, various locations, 1992-Present; Lab Assistant, Streiter Laboratory, Collinsville, IL, 1991-1992; Sales/Customer Service, Sears Roebuck & Company, Fairview Heights, IL, 1989-1991; Fitness/Sales Consultant, Living Well Fitness Center, Belleville, IL, 1987-1989

References Available Upon Request

71—PHARMACEUTICAL SALES

MIKE GRUBBS
12 Hopewell Amwell Road
Hopewell, NJ 08525
(555) 555-1212

QUALIFICATIONS SUMMARY

- Highly accomplished, resourceful, and high-energy **Pharmaceutical Sales Professional** with broad experience including global/national accounts management complemented by extensive account development and cultivation of new markets/territories.
- Track record of sales success characterized by consistent achievement of virtually every sales quota and objective.
- Pragmatic leadership style reflects commitment to long-range strategic planning as well as encourages a collaboration among team players.
- Exceptional abilities in cultivating new business, broadening penetration within accounts, and achieving closure upon completion of highly effective presentations.
- Utilize solutions-oriented sales approach emphasizing customer needs.

PROFESSIONAL EXPERIENCE

1995–Present BOEHRINGER INGELHEIM • Danbury, CT
International Business Development Manager
- Established distribution and marketing strategies for overseas accounts, increasing sales by 85% within six months; successfully opened 21 new accounts.
- Communicated daily with distributors to identify growth opportunities.
- Analyzed sales and provided manufacturing production projections.
- Recruit, direct, and effectively manage staff of 14 Senior Sales professionals.

1991–95 BAYER PHARMACEUTICALS • Milford, CT
National Sales Manager (Promotion, 1993–95)
- Developed a national sales network for new big box accounts; achieved brand prominence in top 10 retailers nationwide.
- Planned territorial expansion and major market sales trips.
- Facilitated financial planning and advertising for retail store management; effectively negotiated sales contracts.
- Between 1993–95, consecutively doubled sales nationally in key product segments.
- Worked closely with scientists/chemists on product development and marketing specifications.

Senior Sales Representative (1991–93)
- Key account responsibility for major accounts throughout the Northeast; achieved 200% of sales objectives in 1991, 1992, and 1993.

1989–91 BAXTER HEALTHCARE MANAGEMENT SYSTEMS • Boston, MA
Sales Representative
- Developed a regional sales network for state-of-the-art records management system; successfully closed 80% of all cold calls and contributed nearly $5MM in incremental sales to bottom line.

EDUCATION CLAREMONT McKENNA COLLEGE • Claremont, CA
Bachelor of Arts (1989, magna cum laude) — **Dual Major: Economics/Literature**

Successfully completed Management Development Program (Bayer, 1991–94)

72—PHYSICAL THERAPIST

BEATRICE McNOLTY

1212 MEADOWCROFT AVENUE, CONNELLSVILLE, PA 15317   **432.555.7802**

PHYSICAL THERAPIST
Orthopedic Applications / Earned Academic Appointments / Knowledge of HICFA PPS
8 Years Experience In Encouraging Independence In Skilled Nursing, Acute Care And Outpatient Environments

SUPPORTIVE EXPERIENCE

PHYSICAL THERAPIST ... THE HEALTH CENTRE, CHICAGO, IL   **8/94 to Present**
Metropolitan Hospital Affiliate - Skilled Nursing Facility And Head Injury Clinic - Staff of 16

Perform and oversee on-site therapy at a skilled nursing facility and a head injury clinic. Handle intake assessments and follow patients through to discharge. Contribute to interdisciplinary short- and long-term goal setting. Identified and implemented quality improvements.

- Work daily with up to 30 residents and patients learning to manage orthopedic (knee, hip and hands), cardiac conditioning, status post-stroke conditions and head/spinal injuries for a 220-bed skilled nursing facility and a head injury clinic.
- Coach patients' maximum participation - - focus them on their achievements versus their limitations and employ a hands-on approach reinforcing their progress and self-esteem.
- Handle on-site visits of residents' living situations prior to discharge. Inspect home surroundings and recommend environmental adaptations. Visit personal care homes on patients' behalf - - orient staff regarding personalized care and special needs.
- Supervise 16 physical therapists and assistants - - establish daily priorities, review results and keep them motivated. Interviewed, trained, precepted and evaluated 18 interns and new hires to date.

ACHIEVEMENTS:

- Led a quality improvement effort boosting resident and family satisfaction rating to 98 percent. Simultaneously instituted a program alleviating pressure via positioning changes.
- Developed a policy and procedures manual used facility-wide that standardized employee orientation and improved accountability by better communicating expectations.
- Created a program simulating patients' living environments for a "trial run" of independent living prior to discharge that enabled therapists to better address residents' limitations for long term planning.
- Instrumental in implementing in-house HICFA Prospective Payment System and establishing guidelines. Remain on-call to perform evaluations.

PHYSICAL THERAPIST ... FLEX PHYSICAL THERAPY, CHICAGO, IL   **1/85 to 8/94**
Promoted From Assistant Upon Acceptance To Masters Program (5/88) - Outpatient Facility

- Initially performed moist heat and stim, vibration and whirlpool therapy for up to 20 patients daily recovering from sports injuries, orthopedic, post-op rehabilitation and stroke-related challenges.
- Completed field experience on-site and went on to lead implementation of work hardening program facilitating workman's compensation patients' return to the workplace - - spearheaded its marketing.

EDUCATION

NORTHWESTERN UNIVERSITY, CHICAGO, IL

Master of Physical Therapy

B.S. Psychology - Minor: Physics - Magna Cum Laude Graduate - GPA: 3.88/4.0
(2-Time Scholarship Recipient / Phi Beta Kappa, National and Psychology Honor Societies)

CONTINUING EDUCATION ... "Current Concepts in Knee Management and Care," Robert Murray, M.Ed., PT, ATC. "Maximum Function Through Cardiopulmonary Intervention," Dan Johnson, PT, RRT

PROFESSIONAL AND COMMUNITY AFFILIATIONS

American Physical Therapy Association (Geriatric and Acute Care Sections). Partners in Education Tutor. Chicago Marathon. Special Olympics. **APPOINTED AND ELECTED POSITIONS ...** Northwestern University School of Physical Therapy Admissions Committee (1997). Green and Gold Society Executive Board.

73—PHYSICAL TRAINER

44 Linton Blvd.
Miami, FL 33333
(305) 555-1711

GREGORY A. HEINES
PERSONAL TRAINER
Specializing in ...   Body Building / Fitness & Health / Competition Training
15+ Years of National & International Experience Supported by Solid Testimonials

****
****
***

HIGHLIGHTS OF AWARDS AND RECOGNITIONS

1st Place World Amateur Championships	1982
1st Place Mr. Pennsylvania Championships	1983
1st Place Russian Grand Prix	1985
1st Place International Over 30 Competition	1997
2nd Place Eastern Pro Amateur	1990
3rd Place Canada Pro Cup	1983
3rd Place European Invitational	1995
4th Place Toronto Invitational	1986
5th Place Joe Weider's Mr. Olympia	1984

HIGHLIGHTS OF EMPLOYMENT

Personal Trainer

Gold's Gym, Reading, PA	1992 - Present
World Gym, Pittsburgh, PA	1988 - 1992
The Downtown Gym, Pittsburgh, PA	1986 - 1988

TRAINING AND EDUCATION

Personally Trained by Gus Johnson and Harry Nordstrom	1980 - 1984
Graduate of Pittsburgh School of Physical Fitness	1989
Graduate of Joe Weider's Physical Fitness Training Program	1991

Professional References and Portfolio Available Upon Request

74—PHYSICIAN

DONNA M. NEWTON, M.D.
"Diplomate American Board of Anesthesiology"

<u>Home Address</u>
18 Anna Court
Vera, IL 60148
(312) 555-7609

<u>Business Address</u>
1801 Spiro Street, #101
Slaven, IL 60145
(312) 555-1099

<u>LICENSES & CERTIFICATION</u>: State of Illinois (#303214), State of Wisconsin (#207772), State of Florida (#36485068)
Board Certified: American Board of Anesthesiology, 4/79 (#9999)

<u>QUALIFICATIONS</u>:

General Anesthesia	Open Heart	Pre-Op Consult
Local	Pain Clinic	Post-Op Consult
Hypothermia	Hypotensive	Inhalation Therapy
Regional Anesthesia	Intravenous	Nerve Blocks
(Spinal, Epidural, Caudal)	Emergency Treatment	

<u>PROFESSIONAL EXPERIENCE</u>:

<u>Anesthesia Associates of Lombard County, P.A.</u>, Lombard, IL   1990 - Present
Practice of Anesthesia / Partner: Ursula Pels Heart Institute
<u>North Broward Medical Center</u>, Chicago, IL   1979 - 1990
Practice of Anesthesia
<u>Medical College of Wisconsin</u>, Milwaukee, Wisconsin   1976 - 1977
Assistant Professor

<u>EDUCATION</u>:

<u>University of Medicine</u>, Zagreb, Yugoslavia
Degree: Medical Doctor: (1966-71)

<u>K.B.C.</u>, Zagreb, Yugoslavia
Internship: (1972-73)

<u>Medical College Of Wisconsin</u>, Milwaukee, WI
Residency
V.A. Hospital / Milwaukee General Hospital

<u>Hospital for Children</u>
Toronto, Canada
Pediatric Anesthesia: 1975

<u>Institute Za Tumore</u>, Zagreb, Yugoslavia
Research on Electro Anesthesia
V.A. Hospital

<u>University of Arkansas</u>
Little Rock, Arkansas
Obstetric Anesthesia

<u>*Continuing Education*</u>:
Open Heart Workshops (AMA Certified, annually, 1987 - Present)
TE Workshops, (Vail, CO, 1989, 1992, 1996)
Meet AMA Yearly Mandates for Continuing Education

<u>MEMBERSHIPS</u>:

- American Medical Association
- Illinois Medical Association

- American Society of Anesthesiologists
- Illinois Society of Anesthesiologists

<u>LIABILITY INSURANCE</u>: Over 17 years of professional practice. Never had a judgement or settlement in any professional liability case. Present coverage: C.N.A., Professional Liability Company

- Professional References & Supporting Documentation Furnished Upon Request -

75—PROGRAM DIRECTOR

Pat McHugh
Park View on the Ridge, #42
Hyde Park, NY 11493
(444) 555-1212

Profile
- **Human Services Program Director** with demonstrated administrative management expertise.
- Excellent communicator with ability to motivate and direct efforts of others; highly methodical approach to responsibilities with outstanding follow-through skills.
- Strong project management/implementation skills complemented by excellent trouble-shooting skills, documentation and technical writing abilities; expertise in conducting comprehensive investigations.
- Special skills in developing individualized treatment programs and person-centered plans.
- Keen sense of personal drive and initiative; dedicated team player.

Education
SKIDMORE COLLEGE • Saratoga Springs, NY
- *Bachelor of Arts, Psychology* (1984)
- Attended Syracuse University • London, England (1983)

Professional Experience

1987–Present PARK HOUSE, INC. • Hyde Park, NY
Residential Program Coordinator (promotion, 1994–Present)

> *Job title in Profile helps reader categorize resume without any interpretations.*
>
> *Thorough description of responsibilities in a position that does not lend itself to many quantifiable accomplishments.*
>
> *Easy-to-read layout is critically appealing.*

- Orchestrate direction and management of programming for approximately 55 individuals living in 8 group homes and supported living programs throughout Greater Hyde Park area; many residents are dually diagnosed with developmental disabilities as well as mental illness.
- Oversee programmatic services, ancillary support services, program staff training, and licensing of Park House, Inc.; develop programming in these key areas as well as volunteer services, community services, and behavioral programs; provide input to staff evaluation process.
- Author and develop highly individualized treatment programs and comprehensive person-centered plans; ensure appropriate implementation. Program objectives incorporate long-term goal of moving residents into supported living as desired by the individual.
- Overall responsibility for quality assurance of agency and compliance with state and federally mandated requirements; interface extensively with wide range of ancillary/support personnel, state agency representatives, family members, and medical healthcare providers.
- Highly effective in role as coach in MBO process being implemented within group homes.
- On day-to-day basis, participate in intake and discharge of residents; assess appropriateness of placement into program, collaborating with DMR representatives, case managers, etc.
- Oversee residential supervisors; participate in selection of paraprofessional staff.
- Act as consultant and facilitate training of staff on organizational and programming issues, particularly with regard to person-centered plans.
- Ensure readiness for full evaluation/licensure process every two years; participate in remedial action plan writing and implementation as well as rewrite of policies as appropriate.

Residential Supervisor (promotion, 1990–93)
- As *Supervisor* of Group Home Residence in Hyde Park, NY, for 6 adults with developmental disabilities, ensured that the physical, developmental, and emotional needs of the residents were provided for by the professional and paraprofessional staff.
- Developed and implemented teaching strategies in conjunction with residential staff; updated on an ongoing basis to allow for optimal success; wrote quarterly review summaries.

Pat McHugh
Page Two

Professional Experience

PARK HOUSE, INC. *(cont'd.)*
Residential Counselor (1987–89)
- As *Counselor,* ensured a safe and therapeutic living environment for residential clients; delivered direct care to residents.
- Provided counsel and instruction to residents in activities of daily living (ADLs).
- Effectively implemented prescribed teaching strategies.
- Interfaced extensively with members of the Management Support Team and other residential staff on client programming issues.
- Maintained appropriate documentation.

1984–86   OXFORD ACADEMY • West Hartford, CT
Self-Contained Classroom Teacher (promotion, 1985–86)
Resource Room Teacher (1984–85)
- Provided instruction to high school students with emotional disturbance and learning disabilities.
- Served as Reading Tutor (1985).

Community/Civic

HYDE PARK PUBLIC HEALTH/MENTAL HEALTH SERVICE • Hyde Park, NY
Board of Directors, Member (1993–Present)

76—PUBLIC RELATIONS

Ariel S. Conroy

212 · 555 · 5555

10 West 25th Street, Apt. B · New York, NY 10001

Event Planning · Public Relations · Media

High-energy, background in fast-paced corporate event planning, promotion, and media relations / production. Possess outstanding cross-industry skills, superior presentation abilities, a passion for excellence, and a contagious enthusiasm. Tenacious and resourceful; will work any hours necessary and will always find a way to get project done on-time / on-budget.

Summary of Qualifications

- Blend creative and administrative abilities to coordinate unique corporate affairs, and media meeting planning for Dun & Bradstreet, Canadian Imperial Bank (CIB), and Jump-Start Productions.

- Manage budgets; select event venues; handle bookings, travel planning, entertainment, and gift selection. Team with design groups to create event ads and collateral materials.

- Function as associate producer on commercials, and as media marketer for Jump-Start Productions, a television / cable commercial production firm. Maintain excellent rapport with producers, clients, and high-profile talent.

- Highly experienced in PC word processing, database / spreadsheet design, and presentation development. Familiar with Mac programs.

Career Highlights

- Helped plan and deliver Dun & Bradstreet's largest and most luxurious special event, a $1 million golf / spa outing at Pebble Beach, CA that was attended by nearly three hundred top clients, executives, and their guests.

- Coordinated cocktail receptions, luncheons, company tours, and interviewing rounds for D&B's recruiting events. Created sophisticated spreadsheets to organize hundreds of participants into six $100 thousand events.

- Planned high-profile golf and entertainment excursions and closing dinners for CIIB. Coordinated cocktails, dinner menus and locations, transportation, executive suite at Madonna concert, and other entertainment. Purchased amenity gifts, inspected sites and paid invoices.

Professional Development

PUBLIC RELATIONS AND EVENTS COORDINATOR Dun & Bradstreet, New York, NY	1996 to present
MEDIA MARKETING REPRESENTATIVE (freelance) Jump-Start Productions, Inc., New York, NY	1994 to present
SPECIAL EVENTS COORDINATOR / PROJECT ASSISTANT Canadian Imperial Bank, New York, NY	1991 to 1995

Education

Bachelor of Arts in Communications, Queens College, Flushing, NY, 1990

Areas of Expertise

corporate representation

PR strategies

press releases

presentations

conflict mediation

investor relations

consumer relations

event coordination

budget development

travel planning

meeting planning

venue selection

cocktail receptions

luncheons / dinners

entertainment selection

golf outings

theme design

invitations

corporate gift selection

collateral materials

vendor payment

Author's favorite.

Excellent use of left column (created in Word tables).

Good example of layout; leading with summary, qualifications, and career highlights, not focusing on where she worked but what she did.

77—PURCHASING MANAGER

MARGARET P. CAMBRIDGE, CPM
8745 St. Michael's Way, Baltimore, Maryland 43121
Email: MPCambridge@hotmail.com
Home: (410) 555-0216

CAREER PROFILE:

- Procurement & Supply Chain Management
- Materials Management & Inventory Control
- Supplier Sourcing & Analysis
- Contract Negotiations & Administration

- Continuous Process Improvement
- Cost Reduction & Containment
- Supplier Relationship Management
- Staff Development & Leadership

Expert planning, organizational, leadership and negotiation skills demonstrated through management of over $150 million in domestic and international contracts throughout career. Innovative in identifying and implementing immediate changes in procurement policies, systems and methodologies to improve performance, capture opportunities, and facilitate positive and profitable change. Excellent communicator and tenacious negotiator with a strong record of reducing costs and enhancing quality of operations. PC proficient with Word, WordPerfect, Excel, Quattro, Internet and Performance Now.

PROFESSIONAL EXPERIENCE:

BUTTE PERFORMANCE MACHINING, INC., Baltimore, Maryland   1984 to Present
(Acquired by Advanced Manufacturing Technologies in January 1998)

Purchasing Manager

Promoted through several increasingly responsible purchasing assignments during a period of rapid growth and market expansion of this $80 million manufacturer with 500 employees. Challenged to link purchasing to corporate vision based on delivering the highest quality product at the lowest cost with the fastest delivery through cost containment, continuous process improvement, technology innovation, supplier relations and team leadership.

Direct the entire purchasing organization accountable for purchases in excess of $40 million annually. Scope of responsibility is expansive and includes procurement of thousands of products from 150+ domestic and international suppliers. Personally manage large dollar purchasing negotiations and major supplier relationships. Recruited, trained and currently direct an eight-person purchasing/administrative staff.

Key Projects & Highlights:

- Played a key role building Butte's purchasing department from start-up to its current $40 million in annual volume. Positioned the purchasing function as a strategic business partner with core operations delivering profit contributions up to 6% annually.

- Achieved strong and sustainable profit improvements through aggressive control of annual purchasing costs. Implemented economic ordering quantity methodologies and JIT purchasing systems. Reduced cycle time on 40% of purchase volume from two weeks to one/two days.

- Spearheaded automation of the entire purchasing function. Worked with IS to create operating infrastructure, automatic supplier stocking programs and corporate-wide procurement card initiative. Significantly improved operating efficiencies, streamlined processes and improved availability/accuracy of key purchasing information.

164

MARGARET P. CAMBRIDGE, CPM – *Page Two*

BUTTE PERFORMANCE MACHINING, INC. (*Continued*):

- Orchestrated the introduction of supplier quality, rating and management programs designed to foster a unique relationship between Butte and its primary suppliers. Instilled a sense of personal commitment by suppliers to the quality and responsiveness of purchased items.

- Designed and led a series of corporate training programs (e.g., cross-training, career development, motivation, incentives) which delivered measurable improvements in employee productivity, morale and job satisfaction. Maintained stable department and ranked as one of the top managers in the corporation based on employee surveys.

- Appointed to corporate ISO 9000 and continuous process improvement taskforces. Led purchasing team through extensive internal audits with no findings.

LIGHTNING INDUSTRIES, Randolph, New York   1980 to 1984
(*Formerly Mitchell Manufacturing Company*)

Customer Service Manager

Led the recruitment, training and leadership of a team of 10 customer service representatives responsible for managing sales, marketing, product information and project management inquiries from clients nationwide. Effectively managed personnel in a fast-paced, customer-driven business environment. Built reputation of department to one of the best in the industry based on responsiveness, professionalism and technical knowledge.

CERTIFICATIONS & EDUCATION:

Certified Purchasing Manager (CPM)
NATIONAL ASSOCIATION OF PURCHASING MANAGERS (NAPM)

General & Business Coursework
STATE UNIVERSITY OF NEW YORK / SYRACUSE UNIVERSITY

Graduate of more than 100 hours of continuing professional education sponsored by NAPM, American Management Association, Karrass, Ernst & Young, Stephen Covey and numerous others. Topics included:

• Sourcing Offshore	• Cost/Price Analysis	• Team Leadership
• Supplier Assessment/Qualification	• Total Inventory Management	• Conflict Resolution
• Effective Negotiating	• ISO 9000	• Employee Relations

PROFESSIONAL AFFILIATIONS:

National Association of Purchasing Managers
Baltimore Association of Purchasing Managers
American Diabetes Association (Past Board Member)
United Way

> *All lowercase letters used for name works for this creative discipline.*
>
> *Up front summary of Career Skills and the talent summary helps the reader quickly understand the skills of the candidate.*

david j west
115 north union boulevard, colorado springs, co 80909 • 719/555-9050

uniquely qualified for . . .

on-air talent - senior producer - public relations - journalism - events management

career skills

- personable, articulate and professional with elements of uniqueness
- long-term on-air/camera and consulting experience
- proven ability to adapt quickly to challenges and changing environments
- excellent public speaker and widely experienced in communications
- comfortable with powerful and celebrity personalities
- appropriate academic credentials plus 'in-the-trenches' experience
- smarter than the average bear (*cum laude* graduate at wake forest university)

formal education

master of arts, journalism and business, new york university
bachelor of arts, communications, wake forest university
graduate, city university, london, u.k.

practical experience

television

prime/fox network, denver - commentator, grandprix of denver
krma-tv, denver (pbs affiliate) - on-camera host
kact-tv, aurora - anchor, reporter & writer
cbs network, new york - production assistant

radio

khih, denver - air talent
ks-104, denver - morning news co-host, air talent & promotions director
khow-fm, denver - afternoon news co-host and air personality
jones satellite network, ac - air talent, heard in 70 cities worldwide
kmji/xl-100, denver - air talent and assistant production director
ktfm, san antonio - morning show co-host, promotions director

production (audio)

tci cable - writer, voice-over talent for various clients and movie channels. commercials, narration, audiotext and interactive telephone applications for hundreds of clients, agencies, production houses and television stations.

references, tapes and further data upon request

79—REAL ESTATE MANAGER

Art Diceman
21 Terra Drive
Laguna Hills, California 92653
(714) 555-3333

Real Estate...Construction Management...Company and Franchise Development
Site Layout / Site Design and Selection / Lease Negotiation / Permitting / Zoning / General Contracting / Bidding / Negotiations / Contractor Relations / Executive Leadership

Vice President of Development for California Pizza Kitchen, a major PepsiCo-owned chain with retail outlets in 19 states. Promoted to higher levels of responsibility over the last 10 years, gaining expertise in all phases of construction, franchise development, and management of real estate for renovation and business expansion in national and international markets. Key management skills include:

Purchasing	Logistics/Strategic Planning	Vendor Relations
Team Building	Customer Service	Sales/Promotions
Training/Development	Cost/Risk Analysis	Budgeting
Market Research	Brand Expansion	Project Planning

A process-oriented leader whose ability to consistently streamline operations has resulted in significant cost savings, increased productivity and business capacity, and million-dollar gains in profitability.

Professional Experience

PEPSICO INC., 1985-Present

California Pizza Kitchen (CPK), Los Angeles, California, 1996-Present
Vice President, Development Services: Oversee the national development strategy for this casual dining restaurant with annual revenues of $180 million. Supervise facilities department that provides service to 77 stores with a $3.5 million repair/maintenance budget and a $4 million capital improvement budget.

Scope of accomplishments:
- Reduced service costs in facilities department from 3.7% of sales to 3.2% in six months, working toward a goal of 2.5%.
- Established the company's first real estate penetration strategy. Currently analyzing demographics in key markets and their respective trade areas to facilitate the future growth of franchises and company stores. Manage staffing, site selection, quality control, site development, and franchisee relations.
- Created and implemented a national construction program using a proven project management approach. Developed a tracking system and convinced the company to leverage the purchasing power of PepsiCo Food Services as an added cost control measure. Sought national contractors that resulted in an immediate annual cost savings of $150,000.
- Launched an effective franchise development strategy to increase and control the growth of new businesses with uniform high-quality standards. Provided comprehensive training from real estate development to initial store opening.
- In conjunction with the CFO, helped lay the groundwork to establish an international presence for CPK in the Pacific Rim.
- Currently involved in the effort to modernize the store's image and expand into new business segments. Designed and implemented a new channel for brand expansion (ASAP) to compete in the express dining market.
- Disposed of excess properties that yielded a net gain of $1.2 million.

79—REAL ESTATE MANAGER (CONT.)

Professional Experience (cont.)

PEPSICO INC., 1985-Present

PepsiCo Food Services (PFS), Irvine, California, 1994-1996

Director of Equipment Marketing and Support, North America: Directed activities for 40 sales, expediter, and project management professionals supporting $120 million procurement and supply business for Taco Bell, Hot n Now, Chevys, and CPK.

- Aligned strategic direction of PFS with that of Taco Bell, KFC, and Pizza Hut. Consolidated non-sales activities from four venues to one, achieving a $3-5 million cost savings through the elimination of redundant functions and satellite offices.
- Implemented multi-functional teams to improve the efficiency of service delivery. Created new delivery strategies and project management positions into the service cycle. Reduced backorders from an average of 10% to less than 2%.
- Positioned sales and marketing teams to focus on franchise development. Instilled project management techniques, expanded licensing capabilities, and implemented extensive skills training.
- Established new business and distribution channels to elevate the company to a full-service organization, resulting in additional revenue of $6 million for PepsiCo and non-PepsiCo concepts.

Taco Bell Corp., Marlton, New Jersey, 1985-1994

Director of Development Services, Northeast and Canada: Supervised 14 construction engineers in the development of a $150 million dollar real estate portfolio. Served as national liaison for equipment procurement and delivery issues. Recruited to reengineer internal delivery processes as well as strategically plan and execute the construction of new restaurants. In 1985, only 9-12 new stores were being built per year in this region and 85-100 stores nationally.

- Increased staff 150% and developed the first national training program for construction managers. Program consisted of multi-level training in development processes, field investigation, and team problem solving, resulting in a more skilled talent base and a newly created pipeline of experienced workers for future projects.
- Spearheaded the rebuilding of a Taco Bell store destroyed during the Los Angeles riots. Store was fully-operational in 48 hours, generating $5-7 million of free advertising and marketing spin.
- Streamlined the entire development process to facilitate unprecedented regional and national growth. By 1994, 125 free-standing Taco Bell stores and an additional 100 Express units had been opened in the same regional geography. Nationally the company approached the capacity to build 800-1000 stores annually, representing an increase of 700-900% with lower costs and higher quality standards.

Previous positions held at Taco Bell Corp. include: **Senior Manager of Construction; Manager of Franchise Development; Construction Manager**

ADDITIONAL EXPERIENCE, 1980-1985

Partner and Vice President of Construction for two New Jersey-based entrepreneurial ventures. Directed all phases of multi-story, mid-rise residential building as well as commercial/light industrial projects including fast food restaurants, warehouse facilities, and strip shopping centers.

Education

Bachelor of Arts Degree in Metropolitan Studies & Architecture
Ramapo College of New Jersey, Mahwah, New Jersey

Lee Wong, GRI   Real Estate Professional

17 Granny Road, Medford NY 11763 ■ voice 516.555.5555 fax 516.555.1111

Broker Associate ■ Relocation Specialist ■ Instructor

- Ten years of comprehensive residential sales experience with Century 21, Prudential Long Island Realty, Coach Realty, and ERA.

- Award-winning performer, consistently placing in the top 10% of Long Island sales force. Perennial multi-million dollar producer.

- Director of Training for ERA. Instructor at American Real Estate School. Participant in Prudential's Mentor Program. Licensed New York State Business School teacher.

- A skillful negotiator knowledgeable in all areas of real estate sales. Expert in residential marketing and promotion. Well versed in National Association of Realtors' code of ethics.

- Utilize sound planning, attention to detail, consultative sales techniques, and company reputation to build customer trust and expedite a property's productive sale.

- Hold over 300 hours of real estate and business training classes, seminars and conferences.

Certification and Licensure

- New York State Licensed Broker Associate

- New York State Licensed Real Estate Agent

- Certified Home Sales Specialist

- Prudential Realty Designated Relocation Specialist

- Licensed New York State Real Estate Instructor

- Licensed New York State Registered Business School Teacher

> *Power heading with title (Real Estate Professional) on same line as his name.*
>
> *Positions the resume to promote certifications, license, and education on an equal par as past experience.*
>
> *Resume successfully indicates his three areas of specialization: Broker Associate, Relocation Specialist, and Instructor.*

Education and Professional Development

- Master of Business Administration, Hofstra University, Uniondale, NY, 1980

- Bachelor of Arts, Adelphi University, Garden City, NY, 1976

- Hold GRI designation; 60 hours at Graduate Realtor Institute, National Association of Realtors

- 45 hour New York State Broker Course

- 45 hour New York State Licensed Sales Agent Course

- Long Island Board of Realtors Educational Conference. Attended four of last five conferences including breakout meetings and key-note addresses.

- Tom Hopkins Sales Seminar and Mike Ferry National Real Estate Sales Seminar

- Floyd Wickman's "Sweathogs" Program, a twelve week class, nuts-and-bolts training course that stresses basic sales skills and re-energizes sales professionals.

81—RECEPTIONIST

SANDRA CUMMINGS
115 North Union Boulevard
Colorado Springs, Colorado 80909
(719) 555-9050

OBJECTIVE

An Office Administrative position as receptionist, switchboard operator, customer service specialist or help desk administrator

EDUCATION

Associate of Applied Sciences, Office Services, Knox College-1997
[Keyboarding, multiple-line telephone answering, 10-key, grammar, spelling, office machine troubleshooting and customer service]

Graduate, Computer Tutor Training Program-1998
[Software training includes all Microsoft Office elements including Word, Excel and Powerpoint plus dBase IV and FoxPro]

Graduate, Palmer High School, Colorado Springs, Colorado-1995

PROFESSIONAL EXPERIENCE

McDonalds Restaurant, Colorado Springs, CO (Summers 1994-Present)
[Part-time position to pay for college]

- Started in the position of Crew Member
- Promoted to the position of Crew Leader
- Promoted to the position of Crew Chief
- Promoted to the position of Shift Manager
- Promoted to the position of Assistant Manager

Current, Inc., Colorado Springs, CO (Christmas seasons 1996-1997)
[Part-time position to pay for college]

Served as Receptionist and PBX operator
- Answered and directed 300± telephone calls daily
- Greeted and assisted visitors and gift shop customers

OFFICE SKILLS

Typing at 75 wpm	10-Key by touch	Customer Service
Billing	Cashiering	Purchasing
Records Management	General Office Machines	WordPerfect

82—RESEARCH ASSISTANT

MICHAEL NOEL

506 ATHENS COURT, LIMA, OH 37895   **783-555-1062**

ENTRY LEVEL RESEARCH ASSISTANT

B.S. Biology / Laboratory And Field Research / Natural Sciences / Environmental Projects
20+ Years Transferable Management, Coordination, Customer Service And Decision Making Experience

DEMONSTRATED STRENGTHS

• Reliability • Goal setting • Prioritizing • Follow up • Critical analysis • Tenacity • Organization •
• Commitment to quality • Communication • Team building • Contributing independently • Attention to detail •

HIGHLIGHTS AND EXPERIENCE

- Detail-oriented individual with demonstrated success in managing diverse priorities to meet deadlines. Strong focus on results. Direct exposure to laboratory research through formal studies.
- Hands-on work with variety of equipment - - spectrophotometer, colorimeter, dissecting microscopes - - broad base exposure to general chemical and biological lab equipment, introductory knowledge of electron microscopes.
- Core coursework affords formal studies in Organic Chemistry I and II, Physiology, Cell Biology, Microbiology, Genetics, Anatomy, Ecology, Microscopy.
- Experimented with basic and advanced life forms. Performed gel electrophoresis, bacterial transformation, enzyme assay, artificial resuscitation of organs and DNA isolation and extraction. Cultured and identified bacteria. Manipulated an e-coli penicillin non-resistant strain to a resistant one using DNA selection and manipulation. Injected various chemicals into a frog's heart and documented effects, artificially sustained it.
- Performed mammological research, classified species using skulls and dental structures. Assisted in a project determining whether the white footed mouse preferred new or older trees using on-campus specimens. Findings indicated that the mouse population was higher in older trees.
- Acquired a keen eye for detail in distinguishing plant characteristics in botanical research, giving equal weight to the plant's native environment in making final determinations.
- Identified plants from surrounding area by observation, comparison and microscopic evaluation. Dissected flowers to ensure correct classification. Results are on record at the University Herbarium.
- 19 years in sales, merchandising and inventory management for 100,000 square foot, 10-department grocery store. Hired and supervised a staff of 35. Worked with in-house promotions and merchandising. Lead negotiator and decision maker with suppliers and advertising representatives. Managed vendor involvement to assure maximum salability of available space. Won 2 awards for merchandising and promotional ideas. Entrusted by owner to take over financial management.
- Computer literate. Familiar with Windows 95, WordPerfect, Microsoft Works, word processing, database and spreadsheet applications.

CUSTOMER SERVICE REPRESENTATIVE... HOLLYWOOD VIDEO, BELLE VERNON, PA   **1996 to 1998**
High Profile Position - General Clientele - Total Service

Pursued A.S. in General Studies and B.S. in Biology, 1992 to 1998.

MANAGER ... THE LARGE MARKET, ERIE, PA   **1973 to 1992**
Lead Decision Maker - Personnel - Inventory - Vendor Management - Won 2 Awards

EDUCATION

EDINBORO UNIVERSITY, EDINBORO, PA
Bachelor of Science in Biology (1998)

ERIE COUNTY COMMUNITY COLLEGE, ERIE, PA
Associate of Science in General Studies - Concentration: Biology, Chemistry, Microbiology (1995)

> *Very good layout with central focus on Career Highlights, which are achievement-oriented.*
>
> *Good review of Expertise, particularly for this discipline.*

NICHOLAS BASTINI

151 W. Passaic Street • Rochelle Park, New Jersey 07662
(201) 555-3772 • Email: xxxxxx@hotmail.com

EXPERIENCED RETAIL PROFESSIONAL seeks a position as an AREA / DISTRICT MANAGER.

AREAS of EXPERTISE:

- Recruiting, training and developing staff to high performance levels.
- Scheduling employees in accordance to customer traffic and demand.
- Displaying eye catching merchandise to increase impulse purchasing and impacting sales.
- Developing and implementing sales goals and evaluating performance of staff.
- Reducing costs to ensure optimum profitability.
- Devising promotional activities and responding to sales trends.
- Overseeing all functions pertaining to operations including sales, adherence to company policy, controlling shrinkage, maintaining inventory levels, customer service and strategic planning.

CAREER HIGHLIGHTS:

- **Drove profit margins 30% in 1997 and 27% in 1996.**
- **Achieved 123% of revenue objectives.**
- **Instrumental in the success of 7 new store launches.**
- **Slashed budget expenditures by 11.7% while gaining enhanced productivity.**
- **Invited as Key Speaker at company's conference in recognition of outstanding performance.**

PROFESSIONAL EXPERIENCE:

DESIGNS, INC.   Secaucus, New Jersey
Store Manager   *1995 - Present*
- Direct daily operations of a 18,000 square foot store generating over $9 million annually.
- Recruit, train and manage a staff of 40 employees to peak levels of performance.
- Develop and implement strategic merchandising plans to ensure revenue objectives are attained.
- Ensure outstanding customer service and develop a loyal client base.
- Initially hired as an Assistant Store Manager and promoted in recognition of superior performance.

STERN'S   Paramus, New Jersey
Department Manager   *1989 - 1995*
- Oversaw departmental functions and provided leadership to staff.
- Addressed customer service issues.
- Interacted with buyers to provide feedback regarding merchandising strategies for greatest sales volume.
- Consistently promoted by senior management to positions with greater responsibility.

EDUCATIONAL BACKGROUND:

RAMAPO COLLEGE of NEW JERSEY • Mahwah, New Jersey
Bachelor of Arts: Political Science/History (double major) • GPA: 3.6

84—SALES ACCOUNT EXECUTIVE

J. WILLIAM PARKER
201 E. Tecumseh Street, Tulsa, Oklahoma 74106 • 555 555-1212

PROFILE

Confident, competent, and achievement-oriented ACCOUNT EXECUTIVE offering personal selling skills enhanced by leadership and refined by a formal education and specialized, sales-related training. Excellent customer service, interpersonal, and motivational skills. Personable, positive attitude. Qualifications include skills in:

- Business Development
- Communications
- Customer Service

- Microsoft Office
- Multi-tasking
- Negotiations

- Scheduling
- Territory Expansion
- Windows 95/NT

PROFESSIONAL EXPERIENCE

WE-TALK-2-U COMMUNICATIONS, Tahlequah, Oklahoma - 1996 to present

A seller of specialized two-way radio communications equipment.

Account Executive reporting to the Sales Manager. Responsible for generating new business, maintaining accounts, expanding territory, and increasing sales. Train account base customers in operation of totally integrated field communications system that includes an integrated, two-way digital radio, digital cellular, paging, and voice mail in one unit. Service existing accounts.

- Consistent record of exceeding established sales quotas by 25% +.
- Received Top Producer Sales Award, 1994.

WEAR BEST UNIFORMS, INC., Tulsa, Oklahoma - 1993 to 1996

A nation-wide manufacturer specializing in uniforms and custom apparel for OEM manufacturers, restaurants, service industries, and hi-tech industries that require specialty garments.

Sales Representative reporting to the Sales Manager. Scope of responsibility is focused on development of new business, expansion of territory, generating sales, and increasing profits. Primary emphasis was on building corporate apparel and image program for new clients.

- Averaged one new client account per month in a highly competitive market with extremely high turnover ratio.
- Closed two major accounts projected to generate an additional $200K +.

PRINT-IT-RITE PUBLISHING, Cincinnati, Ohio -1990 to 1993

A religious book publisher specializing in selling the NIV Bible.

District Sales Manager reporting to the National Sales Manager for the Ohio based division of Harper-Collins. Supervised a staff of 2 in-house sales persons and 2 customer service representative engaged in sales activities. Oriented and trained new sales representatives. Managed inside sales, credit and customer support services.

- Increased sales production by 300% within 24 months.
- Significantly increased annual sales production from $2.5 to $5+ million within 36 months.
- Received Top Salesman Award, 1990, 1991, 1992, and 1993.

EDUCATION

Southern Methodist University, Dallas, Texas
  Graduate Studies - Concentration: Sales and Marketing

Christian Heritage College, El Cajon, California
  Bachelor of Science - Major: Business Administration

PROFESSIONAL DEVELOPMENT

Xerox Corporation, Grand Rapids, Michigan
- Professional Selling Skills

Stephen Covey Seminars, Grand Rapids, Michigan
- Stephen Covey Sales Seminars

Professional Experience bullets offer accomplishments, most of which are quantifiable.

Title does a good job of describing the candidate's background/discipline.

FRANCINE B. ROMANO

150 New England Drive
Garden Grove, New York 10576
(516) 555 -5645

SALES & MARKETING PROFESSIONAL

Specialty, Catalog & High-End Retail Accounts Nationwide
Private Label, Branded & Licensed Products

More than 10 years' professional experience planning and managing high-volume regional and national sales programs. Delivered strong and sustainable revenue growth. Qualifications include:

- Strategic Market Planning
- Sales & Marketing Leadership
- New Business Development
- Sales Training & Development
- New Product Introduction

- Key Account Management
- Retail Merchandising
- Vendor Negotiations
- Inventory Planning & Analysis
- Product Incentives & Promotions

PROFESSIONAL EXPERIENCE

National Sales Manager, D'ITALIA, INC., New York, New York (1996 to Present)

Senior Sales Manager with this $125 million global manufacturer of scarves. Challenged to plan and orchestrate an aggressive market expansion for three licenses, and launch the company's entry into private label and non-traditional market sectors throughout the US. Scope of responsibility includes strategic planning, competitive assessment, market positioning, business development, new product introductions and client relationship management. Direct a team of seven.

- Led the successful national market launch of three product lines. Met first year projections and currently on track to achieve 15% increase for 1998.
- Built and managed key account relationships with major retailers, catalog companies and specialty stores including Bergdorfs, Bloomingdales, Macy's, Neiman Marcus and Saks.
- Worked with home office in Italy to facilitate the design, test marketing and US introduction of an upscale private label line. Achieved immediate market penetration with projections for $25 million in first year revenues.
- Developed strategy to leverage D'Italia's position within non-traditional business sectors. Negotiated contracts for the NHL, Collegiate Licensing Committee and several others.

Retail Manager, SAKINAS, INC. (1988 to 1995)

Recruited to this upscale, national retail chain to manage daily operations for start-up and high-growth retail sites throughout the Midwest. Scope of responsibility was diverse and included daily operations management, recruitment, training, scheduling, inventory control, administration and the entire sales, marketing and customer service function. Led a staff of up to 65.

- Led the Chicago store to ranking as the highest volume operation nationwide.
- Managed the start-up of Detroit store. Recruited 45 personnel, created merchandising displays and coordinated grand opening activities. Built operation to solid first year revenues.

EDUCATION

B.S., Marketing, Syracuse University, New York, 1988

86—SENIOR MANAGER

ELIZABETH R. FAIRCHILD

13803 13th Street NE, #1617
St. Paul, Minnesota 55232

Home: (606) 555-0234
Office: (714) 555-0982

MANAGEMENT PROFILE

Distinguished management career developing business systems, processes and organizational infrastructures that have improved productivity, increased efficiency, enhanced quality and strengthened financial results. Expertise in identifying and capitalizing on opportunities to enhance corporate image, expand market penetration and build strong operations. Broad-based general management, financial management and project management qualifications. Outstanding record in personnel training, development and leadership.

- Strategic Planning & Tactical Execution
- Business & Performance Reengineering
- Corporate Culture & Organizational Development
- Service Design & Delivery Systems

- Productivity & Efficiency Improvement
- Leadership Development & Career Pathing
- Consensus Building & Cross-Functional Relations
- Customer-Driven Management

PROFESSIONAL EXPERIENCE

NORTHWEST AIRLINES, St. Paul, Minnesota   1977 to Present

Fast-track promotion throughout 20+ year tenure transcending from field to corporate operations in both start-up and large-scale business locations. Built successful business partnerships, managed cross-functional communications, and designed/implemented proactive organizational development, employee performance and corporate culture programs. Key projects and achievements have included:

Start-Up & High-Growth Operations Management

- Held full decision-making responsibility for the daily operations of the St. Paul facility, Northwest's largest center. Led the operation through a period of significant growth, market expansion and diversification including several mergers, divestitures and volume increases.
- Introduced cost management, conflict resolution and corporate culture change initiatives for the start-up Yellowstone operation (1000 personnel and 65 flights daily). Resolved long-standing communication issues, streamlined systems, and created a highly-successful and profitable operation.
- Planned construction of 15,500 space parking facility for the St. Paul operation. Worked with city officials, designers, contractors, company personnel and service providers throughout project cycle.

Business Process Reengineering

- Orchestrated a complete reengineering of the Ramp Tower operation supporting seven major facilities nationwide. Designed/implemented programs to streamline processes, increase performance and decision-making authority, and position the operation as a cooperative business partner with core operations. Program is currently being implemented throughout Northwest's U.S. operations.
- Spearheaded complete automation of all administrative functions for 6500-employee St. Paul operation. Significantly improved the timeliness and accuracy of key operating, customer and financial information.

Employee Development, Communications & Liaison Affairs

- Created and currently facilitating a corporate-wide conflict management course for 1500 front-line managers in the Airport Customer Service Division.
- Authored a comprehensive operations manual addressing management practices, leadership development, employee evaluation and numerous other organizational issues. Manual was adopted as the corporate standard for all service operations.
- Acted as direct liaison between in-flight services, flight operations, maintenance and airport personnel to build positive working relationships and facilitate cross-functional communications.
- Led integration of more than 150 personnel, all systems and processes into the St. Paul station following Northwest's acquisition of Midwestern Airlines. Completed transition in less than six months with virtually no interruption to service.
- Introduced a series of training programs for over 1500 new and existing personnel. Achieved measurable improvements in employee productivity, operational performance and customer service/satisfaction.

ELIZABETH R. FAIRCHILD

Security, Safety & Inspection Systems, Emergency Preparedness & Crisis Management

- Appointed by corporate committee, following significant public and government scrutiny of the airline industry, to design an emergency preparedness program encompassing more than 600 flights per day and 60% of total St. Paul airport traffic. Program was unanimously approved by Northwest executive management, the FAA, and the Cities of St. Paul and Minneapolis.
- Created a unique program to effectively manage operations during periods of volatility due to both major and minor incidents/delays (e.g., crashes, inclement weather, mechanical problems). Provided employees with a strategic guideline for effective management of customer service and operations. Greatly reduced the number of customer complaints and costs incurred due to cancelled/missed flights.
- Worked with the Director of the St. Paul operation to design and implement a safety and inspection process. Achieved the lowest number of FAA non-compliance findings in the entire corporation.

Career Progression

Northwest Express Yellowstone Manager (1997 to Present)
St. Paul Ramp Tower Manager (1996 to 1997)
Duty Manager (1995 to 1996)
Station Analyst (1995)
Supervisor / Ramp Tower Coordinator (1988 to 1995)
Lead Customer Services Agent (1984 to 1988)
Senior Customer Services Agent (1979 to 1984)
Revenue Accounting Clerk (1978 to 1979)
Flight Attendant (1977 to 1978)

EDUCATION

UNIVERSITY OF MINNESOTA

Candidate for Doctoral Degree in Community Psychology, 1995 to Present
Emphasis in Organizational, Sociological & Psychodynamic Theory/Application

Bachelor of Science in Psychology & Sociology, 1995
*Magna Cum Laude Graduate; Phi Kappa Phi Honor Society; Golden Key National Honor Society
Outstanding Scholarship Award; State College Scholastic Achievement Award*

PERSONAL PROFILE

Professional Affiliations

Northwest Airlines Corporate Mentoring Program
International Airline Women's Association
Society for Community Research & Action
Northwest Airlines Station Manager's Association

Community Affairs

Twin Cities Memorial Hospital Rape Counseling Center
Habitat for Humanity

Research/Publications

Employee Absenteeism
On-The-Job Injuries & Accidents

87—SENIOR SALES MANAGER

SIMON LEWIS

23541 Old Towne Lane
Westin, Ontario L5P 8B3

Phone: (905) 555 -3321
Email: lewis4@aol.com

SENIOR SALES & MARKETING EXECUTIVE
Start-Up & Emerging Technology Ventures

Top-Producing Management Executive with more than 15 years experience in the development, commercialization and market launch of leading edge technologies worldwide. Combined expertise in strategic planning, P&L management, marketing, tactical sales and client relationship management. Outstanding record of achievement in solutions selling and complex contract negotiations. Diverse industry and multichannel sales experience within the financial services, banking, healthcare, retail and Fortune 500 market sectors.

PROFESSIONAL EXPERIENCE:

Vice President of International Sales   1993 to Present
INTEGRATED SYSTEMS WORLDWIDE (ISW), Toronto, Ontario

Technology: E-commerce & OLTP Applications Delivered on Client/Server, UNIX-Based Open Systems

Recruited as Management Consultant to this venture-funded start-up seeking to launch newly-acquired technology from major U.S. manufacturer. Created corporate vision, strategies and tactical action plans to design/launch a complete platform of network and systems management tools to drive the industry toward integrated enterprise-wide/open system solutions. Identified potential markets, led road show presentations and built the organization from concept into full-scale operation.

Accepted permanent position as the Senior Sales & Marketing Executive challenged to build a global sales organization, expand product offerings and capture emerging technology opportunities throughout the healthcare and financial services industries. Hold full P&L responsibility for strategic planning, new business development, technology development and licensing, and technical service/support. Travel throughout North America, Latin America, Europe, Pacific Rim and South Africa.

- Built ISW's entire network of direct sales and multichannel distributors producing $23 million in annual revenues.

- Delivered annual growth averaging 100% in a highly competitive global market. Played key role in a successful IPO in 1993.

- Championed the development and spearheaded the market launch of several new technologies with cumulative revenues of $2 million annually.

- Guided sales team through several complex contract negotiations and closings. Personally closed $10 million contract in Argentina, $3+ million contract in New Zealand and $4 million contract in France.

Director of Sales   1991 to 1993
TECHNOLOGY SOLUTIONS, INC., Cherry Hill, New Jersey

Technology: HP Systems & Customized Applications Distribution & Service

Member of an eight-person senior management team challenged to launch the start-up of a full-scale distribution, maintenance and service operation as part of joint venture between Hewlett Packard and Techx3. Authored strategic and tactical marketing plans to expand market presence and accelerate revenue growth. Established operating infrastructure, led product development, and recruited/led six-person sales team.

- Played a key role in the company's growth from start-up to $14 million in revenues within two years. Personally generated 50% of total company revenues.

- Spearheaded development and market launch of six new product lines that contributed to $4 million new revenue stream.

- Negotiated a highly-profitable corporate licensing agreement with the Bank of New York.

SIMON LEWIS   Page Two

Senior Account Executive   1990 to 1991
TECHX3, Stamford, Connecticut

Technology: Base24, POS, ATM, Financial Management Applications

Planned and orchestrated an aggressive market expansion to capture emerging market opportunities in the retail and banking industries. Scope of responsibility was diverse and included strategic planning, new business development, new product development and client relationship management.

• Built exclusive partnerships with six of the nine largest financial institutions in the U.S. and facilitated the installation/maintenance of the world's leading retail banking software applications. Delivered revenues in excess of $15 million.

Senior Account Manager – Commercial Systems Group   1988 to 1990
HEWLETT PACKARD CANADA, Toronto, Ontario

Technology: Fault-Tolerant Mainframe Class OLTP Processors/Servers & Peripherals

Held full responsibility for sales, marketing and account management programs throughout Canada. Personally managed a highly-complex and lengthy sales cycle. Focused efforts on building key client relationships, designing/implementing solutions-based sales strategies, and positioning HP as the market leader in non-stop computing and back-end processing technologies.

• Spearheaded the launch of several new product lines targeted to major retail customers. Structured, negotiated and closed two major accounts with total value in excess of $4.8 million.

EARLY SALES CAREER   1981 to 1988

Advanced through increasingly responsible field sales positions with **MXT Corporation** and **IBM**. Marketed advanced technology solutions (e.g., data entry systems, POS terminals, financial management systems) to small, mid-size and Fortune 500 accounts throughout the U.S. and Canada.

• Consistently exceeded all corporate objectives and ranked as a top revenue producer.

• Appointed Sales Trainer at IBM's New York City training facility.

CONSULTING EXPERIENCE:

Consultant   1993 to 1994
SAT, INC., Toronto, Ontario

Launched the start-up of an exclusive management consulting firm specializing in the development of sales and marketing strategies to leverage advanced system/network technologies within the retail and banking industries. Worked one-on-one with senior executives of client companies to assess existing needs and develop strategic technology solutions in debit card integration, payment systems support, point of sale automated management facility (PAMF).

EDUCATION:

B.S., Marketing, 1980
RUTGERS UNIVERSITY

88—SHIPPING MANAGER

STEPHAN JUZWIACK

223-47 38 Avenue, Flushing, New York 11358

EXPERIENCED SHIPPING MANAGER

Fourteen years of hands-on experience in effective operation of warehouse shipping department processing $10 million in annual apparel orders.

- Accurate, responsible, and exceptionally alert to compliance standards and errors.

- Work effectively and independently in challenging environments.

- Meet goals with limited staff and resources.

ACCOMPLISHMENTS

Direct order fulfillment, processing and shipping for high-volume, high-maintenance accounts: J.C. Penny, $2 million; Sears, $150 thousand; Filenes, $100 thousand; and Arizona Mail Order, $500 thousand.

Have saved over $100 thousand in chargebacks by comprehensive review of vendor requirements and by strict control of adherence to these requirements.

Consistently maintain outstanding on-time shipping record with very low pre-date or post-date penalty and bounce back ratio.

Developed and sustain productive rapport with dispatchers — Roadway, Yellow Freight, Consolidated Freight, Preston, APA, Red Star, and Gale. Receive prime pick-up times, even on short notice, or for same day.

EMPLOYMENT

Fibrele/Tempo Apparel, formerly Tempo Knitting Mills, New York, NY, 1983 - present
- **Shipping Manager, Fibrele/Tempo Apparel, 1995 - present**
- **Shipping Manager, Tempo Knitting Mills, 1983 - 1990**

Direct all shipping functions for apparel orders shipped via contracted and customer own trucks, UPS or Fed EX, with prime season volume of $50 thousand to $75 thousand in daily shipments of up to 150 cartons.

Inspect all orders for compliance in areas of bar-coding, ticketing, routing (consolidation or separate stores) special labeling and packaging. Generate complicated manifests for multiple distribution points, identify cartons with specific stickers, deliver special UPS/truck instructions. Prioritize shipments by date, review truck schedules and schedule pick-ups. Train new hires and existing staff in all systems, vendor requirements and compliance standards.

Responsibilities expanded, but staff not increased after merger; currently process and ship double volume with same pre-merger UPS shipping clerk and general shipping clerk. Work any hours necessary to get the job done; put in 50+ hour, 6 day weeks from Labor Day to Thanksgiving.

89—SOCIAL WORKER

Elizabeth J. Holmes, LSW
97034 Chrysanthemum Court
Denton, Texas 76201
940-555-5093

Licensed Social Worker

Social Services Professional with well-built community network. Work through Municipal Court, city police, other city departments and non-profit organizations to provide alternative sentencing for juvenile offenders. Goal is to provide more individualized judgments to meet the needs of the juvenile while holding the juvenile accountable for his/her actions.

Education

Bachelor of Social Work

University of North Texas - Denton, Texas

Professional Credentials

Licensed Social Worker - Certification #15851

Excellent lead of career title and summary of experience.
Education listed second because in this environment, education and professional credentials are important.
Very clean and easy-to-read layout.

Employment History

Irving Municipal Court   Irving, Texas
Director of Social Services for the Municipal Court   1996 to Present

Secure community service for defendants throughout the city in cooperation with the Municipal Judge, city departments, non-profit organizations and school districts. Resolve issues and provide information within the scope of the Alternative Sentencing Program. Assist bailiff with Court dockets; schedule, coordinate and attend jail and juvenile (non-traffic) dockets. Build community support and goodwill.

- Developed and expanded Juvenile Alternative Sentencing Program to provide a variety of programs and services for the defendants.
- Secure work providers within the city, with non-profit agencies and school districts. Update provider's need for workers.
- Develop, maintain and monitor community service worksite placements.

Communities in Schools, Denton County   Denton, Texas
Campus Manager   1994 to 1996

Supervise and facilitate school site programs. Work in partnership with school staff, community agencies, parents/families, and students to assure coordination of services for the benefit of the students and their families.

- Drive corporate and community involvement.
- Solicit funding and sponsorships.
- Provide crisis intervention.

Metrocrest Service Center   Irving, Texas
Social Worker   1988 to 1994

Interview and qualify clients in need of financial aid and commodity assistance. Provide information and referrals. Coordinate employment program with community employers and resource services.

90—STUDENT

MARK LUDWIG

517-555-2491

RÉSUMÉ OF QUALIFICATIONS
328 WESTERN MILL
OKEMOS, MICHIGAN 48821

SUMMARY OF QUALIFICATIONS

> *Very clean-looking resume that appropriately highlights education for this recent graduate.*
>
> *Students should do what they can to exploit things like academic honors, achievements, etc., even if it may seem trivial at the time. Employers do pay attention to those intangibles.*

EDUCATION

Master of Business Administration, (expected 1999)
Central Michigan University, Midland, Michigan
GPA 3.0/4.0

Bachelor of Arts in Communications, 1996
Central Michigan University, Midland, Michigan
GPA 3.4/4.0

HONORS/AWARDS

♦High School Valedictorian
♦Won citywide Essay Competition senior year in high school.
♦Won "Best Freshman Essay" Contest out of 800 students at Armstrong State College.
♦Invited to statewide Academic Recognition Ceremony at State Capitol.
♦Awarded distinguished *Silver A* award from Armstrong State College.

SKILLS/STRENGTHS

♦ Strong understanding of financial markets and market development
♦ Proficient with all Microsoft Windows based programs
♦ Extensive background with performing strategic analysis of corporations
♦ Developed multiple presentations outlining strategic recommendations
♦ Proven tact and diplomacy in handling interpersonal relationships

VOLUNTEER ACTIVITIES AND PREVIOUS WORK EXPERIENCE

Merrill Lynch, Lansing, Michigan
Intern (Summer 1997, 1998)
Performed financial analysis of prospective investment opportunities and worked directly with clients to support account manager

SIGMA ALPHA MU FRATERNITY, Central Michigan University, MI
Kitchen Steward (Academic Season 1996, 1997, 1998)
Established kitchen procedures, many still in use, for newly chartered chapter of this fraternity. Responsible for food budget, purchasing food and supplies, interviewing and hiring kitchen personnel, supervising kitchen and dining room operations, and preparing food. Also held positions of Scholarship Chairman and Fundraising Chairman.

SPECIAL INTERESTS

Arts and crafts, badminton and beginning golf.

91—STUDENT

Bethani Blair
15 North Mill Street
Nyack, New York 10960
(914) 555-3160

EDUCATION

Fordham University Graduate School of Social Work - Bronx, New York
Currently Pursuing Masters Degree in Social Work
Expected date of Graduation Spring 1999

Dominican College - Blauvelt, New York
Bachelor Degree in Social Work - 1997

EXPERIENCE

Stony Lodge Hospital - Briarcliff Manor, New York
Therapist - 1995/Present
Provide individual, group and family therapy to adults, children and adolescent patients at this 40 bed resident treatment facility. Develop clinical reviews for managed care companies and maintain extensive contacts with referral sources and families to coordinate patient care planning.
Accomplishments:
- Designed and incorporated educational groups for adolescents dealing with HIV/AIDS, sexually transmitted diseases and alcohol/drug education groups.
- Prepared and presented a series of seminars on teenage suicide predictors and signs of depression for the New York Juvenile Officers Association, Rye High School and Stony Lodge Hospital in-service training.
- Developed and presented a seminar for a local bereavement group regarding depression and psychopharmacological drugs.

Pequannock Valley Mental Health Center - Pompton Plains, New Jersey
Inquiry Worker 1994/1995
Responded to all telephone and walk-in inquires for this out-patient crisis center. Served as member of interdisciplinary treatment team for the purpose of assessing client's strengths and needs. Implemented individual behavioral treatment plans.

REFERENCES WILL BE FURNISHED UPON REQUEST

92—STUDENT

Weston J. Hoffmann
30 Devonshire Place
New Haven, CT 06222
(888) 555-1212

FINANCE PROFESSIONAL
Relationship Management Skills • Complementary Sales Abilities
Organizational Development • Leadership Qualifications
IBM/Mac Expertise (Excel, Word, Windows)

EDUCATION UNIVERSITY OF MASSACHUSETTS • Amherst, MA
• ***Bachelor of Science, Finance*** *(1998)*

CERTIFICATE DES ETUDES FRANCAIS • Paris, France
• ***Degree in French Studies*** *(1997)*
• Successfully completed year-long program (UMASS-accredited)

PROFESSIONAL EXPERIENCE

1995–Present THE CAR STORE OF CONNECTICUT • New Haven, CT
Assistant Operations Manager *(part-time, 1997–Present)*
• Provide financial support and consulting to one of state's largest independently owned/operated cellular/digital telephone stores through accurate budgeting, forecasting, and reporting of financial analysis.
• Develop annual budgets, including hardware, manpower, and expenses.

Excellent use of listing skills first, followed by real strengths for a student—the education background.

Resume is kept to one page. Few student resumes should exceed one page.

Good use of integrating accomplishments into Professional Experience.

Service/Installation Specialist *(part-time, 1995–97)*
• Professionally handled customer installations, performing all service functions for growing customer base of 1,800 (80% individual, 20% corporate accounts).
• Directly responsible for sales and marketing efforts to individual prospects; contributed $2K–$4K/month in new sales installations and contracts (summers/winter breaks while in college).
• Handled all facets of employee-customer relationship and effective resolution of problems.

1993–95 CALDOR • Orange, CT
Accounting Associate
• Assisted with management of Accounting Department, training new personnel on electronic processing systems.
• Handled bank deposit verification, conducted inventory control, and processed purchase orders.
• Recognized for exemplary performance through award of five "Employee of the Month" Certificates.
• Seasonal full-time/part-time position throughout high school.

93—STUDENT/ENVIRONMENTAL ENGINEERING

CAROL LYNCH
12 Granite Avenue
Broomfield, CO 80021

COLLEGE GRADUATE
SEEKING POSITION IN ENVIRONMENTAL ENGINEERING

EDUCATION

University of New Hampshire, Durham, NH
Masters in Science: Environmental Engineering (Marine/Aquatic Engineering), 1998
* Member: UNH Environmental Club, 1997-98
* Member: UNH Campus & Community Clean-up Task Force, 1997-98

Boston University, Boston, MA
Bachelor of Science: Biology, 1996
* Member: BU Students for a Clean Harbor, 1993-96

INTERNSHIPS - Undergraduate and Graduate

ENVIRONMENTAL ASSOCIATES, INC., Dover, NH   1997 and 1998
Environmental Engineering Assistant
• Worked with clients in developing golf courses in environmentally sensitive areas
• Liaison between senior environmental engineer and EPA, State EPA, and other regulatory agencies
• Researched and troubleshot actual and potential environmental problems

JOHNSON AND CAGE, INC., Newington, NH   1995 & 1996
Environmental Engineering Assistant
• Worked with clients in developing office buildings in environmentally sensitive areas
• Performed extensive research and prepared in-depth environment feasibility report based on findings
• Awarded "Most Valuable Intern," 1996 from a group of 16 interns

EMPLOYMENT WHILE FINANCING EDUCATION

Pizza Delivery (20 hours a week), Ginos Pizza, Durham, NH   1997 - 1998
Waiter/Host (20 hours a week), Boston Ale, Brookline, MA   1993 - 1996
Usher - Weekends Red Sox Baseball Games, Fenway Park, Boston, MA   1993 - 1996

References Furnished Upon Request

94—STUDENT/MARKETING

Randy Zanassi
1135 Michigan Avenue
East Lansing, MI 48823
·(517) 555-0588

> *This is a great student resume because it is simple yet articulates the summary of qualifications, education, and a narrative on endeavors prior to entering the professional workforce.*

QUALIFICATIONS

- Marketing Degree from Michigan State University
- Tom Hopkins Seminar Attendee
- Two years experience selling telephone service and publishing materials
- High level of ambition to begin career

EDUCATION

Michigan State University
East Lansing, Michigan   March 1999
Earned a bachelor of arts in Marketing in under the prescribed four year course schedule, while financing my own education. My final two years I was totally self-supportive, working an average of thirty hours per week.

Tom Hopkins Seminar
Detroit, Michigan   February 10, 1999
The seminar "How to Master the Art of Selling" shall improve my inherent sales abilities. Learning various personnel skills for applicable situations will be an invaluable asset to my career.

WORK EXPERIENCE

United Parcel Service, Lansing, Michigan   November 1997 to present
Working at UPS enabled me to earn enough to support myself in school. I earned over $12,000 per year, an impressive accomplishment for a college student. As well, I maintained the highest production average at our Center.

Sprint Telephone Division, Lansing, Michigan   1996
Sold local telephone feature services to the consumer market in Lansing. This experience paved the way for my future career path in sales. After three months at Sprint, I was the sales leader among the part-time college students and enjoyed the interaction with the customer. The only reason I left was to move on to UPS, where the part-time earning potential was greater.

American Collegiate Marketing, Lansing, Michigan   1995
This position was my introduction to sales, where I worked in a call center selling magazine subscriptions. Though I enjoyed the sales environment, my desire was to sell closer to customer in a more direct manner.

SKILLS / INTERESTS

Beyond my formal education, I have a working knowledge of MS Word, Excel, Outlook and Explorer. I also enjoy playing golf, tennis and fishing.

Easy-to-read format leading off with her experience and accomplishments. She took more of a narrative approach in describing the Professional Experience.

LORI PIORRIZI

151 W. Passaic Street • Rochelle Park, New Jersey 07662 • **(201) 555-3772**

SPECIAL EDUCATION TEACHER

- Experienced in successfully working with Special Education students and developing skills in students at all levels of achievement.
- Utilize creative skills to design and implement well-received lesson plans and program structure.
- Establish learning environments which meet the physical, emotional, intellectual, social and creative needs of children.
- Create yearly course work including the selection of teaching materials.
- Effectively counsel students and parents on goals, objectives and plans.
- Rapidly develop and adjust lesson plans to meet unforeseen classroom situations.

SELECTED ACCOMPLISHMENTS

- **Significantly increased enrollment and student learning by implementing innovative programs.**
- **Instrumental in State accreditation.**
- **Implemented a revised reading program to interest young children.**
- **Planned numerous extracurricular activities.**
- **Recipient of numerous letters of appreciation from parents for dedication and effort.**

PROFESSIONAL EXPERIENCE

The Windsor School - Pompton Lakes, New Jersey   1989-1998

Special Education Teacher (1997-1998)

Administered curriculum for 5th and 6th grade Special Education students (emotionally disturbed, learning disabled and neurologically impaired). Developed lesson plans and instructed all major subject areas including reading, grammar, science and social studies. Assessed student abilities and evaluated performance; conducted parent-teacher conferences to provide parents with student development reports. Counseled students and parents to resolve learning and discipline problems. Developed monthly newsletters and participated in IEP meetings for each child.

Elementary Teacher (1989-1997)

Taught 4th grade for 3 years and full day kindergarten for 4 years. Directed the pre-school and junior kindergarten program for 1 year. Revised curriculum and reading program for kindergarten class. Generated progress reports and evaluated students through report cards. Authored monthly newsletters. Created lesson plans and developed learning centers. Accountable for kindergarten screening and direction/production of annual holiday pageant.

EDUCATION / CERTIFICATION

- *Master Degree in Special Education* (currently pursuing, 12 credits completed)
- *B.S. in Elementary Education* (*Concentration in Early Childhood*) • 1988
  Jersey City State College - Jersey City, New Jersey

- *New Jersey Certificate in Elementary Education (K-8)*
- *New Jersey Certificate in Special Education (pending)*

96—TEACHER

Because she has remained a teacher her entire career, using a functional layout highlighting Core Competencies works very well.

ALEXANDRA N. WINSLOW
8762 South Washington Avenue
Carmel, Indiana 47663
(717) 555-8776

CAREER SUMMARY:

Dedicated **EDUCATOR** with 20+ years of teaching experience. Recognized for innovation in program development, instruction and administration to meet the needs of a broad range of students. Effective communicator, writer, administrator and student advisor/advocate.

CORE COMPETENCIES:

Curriculum Development & Instruction

- Assisted in the development, validation and enhancement of curricula. Introduced hands-on tools (e.g., computer technology, outside classroom activities) to improve classroom interest and retention.
- Taught a full academic curriculum (e.g., reading, writing, communications, mathematics, social science) to children ages seven to 13.
- Designed and implemented customized teaching programs to allow emotionally, physically and learning disabled students to be mainstreamed into the classroom.
- Launched a highly-successful peer tutoring program designed to improve interaction between upper level and younger students while fostering communication and mentoring skills at all levels.

Administration & Special Activities

- Appointed Vice Principal with responsibility for a diversity of functions including faculty recruitment and scheduling, curriculum development, discipline, parent/community affairs, and special events planning.
- Prepared documentation for recertification by the National Accreditation of Schools Committee as a member of a cross-functional faculty/administration committee.
- Provided classroom training, performance evaluation and motivation as a mentor to student teachers completing college requirements for an education degree.

Community & Public Relations

- Built partnerships with local companies and developed a series of seminars to introduce junior high school students to the business world. Coordinated speaker selection and topics of discussion.
- Participated on cross-functional teams of educators and administrators to design programs to improve the quality of educational curricula, enhance parent and community relations, and increase student participation both inside and outside of the classroom.

PROFESSIONAL EXPERIENCE:

Teacher / Administrator	ST. MARY'S DAY SCHOOL, Carmel, IN	1990 to 1998
Elementary Education Teacher	INDIANA PUBLIC SCHOOL SYSTEM	1982 to 1989
Elementary Education Teacher	MICHIGAN PUBLIC SCHOOL SYSTEM	1977 to 1981

EDUCATION:

B.A., Education (Minor in History), MICHIGAN STATE UNIVERSITY, 1977

Permanent Certification in Elementary Education (K-8), Indiana
Past Certification in Elementary Education (K-8), Michigan

Graduate of 100+ hours of continuing professional education and graduate studies sponsored by the College of the Midwest, Ball State University, Purdue University and Michigan State University.

Good use of title at head of the resume. This candidate is just finishing school, so listing education at the top is appropriate.

Richard Jesonowski

115 North Union Boulevard
Colorado Springs, Colorado 80909
(719) 555-9050

Telemarketing Specialist - Verifier - Team Leader

- Qualified by practical experience in Telemarketing, most recently as a Verifier
- Proven ability to lead sales teams in fast-paced and high-volume environments
- Able to coordinate multiple projects and meet deadlines under pressure
- Outstanding record in training, motivating and retaining employees
- Intimately familiar with telemarketing business methods and applicable laws

EDUCATION AND TRAINING

Currently in final year of a Bachelor of Arts program, Marketing, University of Colorado
Associate of Arts, Psychology, Pikes Peak Community College, December 1993
Graduate, Dale Carnegie Sales Training Program, July 1995
Graduate, Karrass Negotiation Seminar, August 1996

PROFESSIONAL EXPERIENCE

January 1994
to
Present

VERIFIER - TEAM LEADER - TELEPHONE SALES REPRESENTATIVE
MATRIXX MARKETING, COLORADO SPRINGS, COLORADO

Started in the position of Telemarketing Sales Representative (01/94 to 06/94):
    Assigned a wide variety of products and services to sell, including
    Credit Cards and Long-Distance Telephone Services.
Accomplishments:
- Consistently met or exceeded assigned sales goals
- Regularly the top producer within assigned team
- Unique record of maintaining a 100% level of verified transactions

Promoted to the position of Team Leader (07/94 to Present):
    Responsible for the motivation and training of 15± assigned Telemarketing
    Sales Representatives.
Accomplishments:
- Reduced team turnover from 312% to 22%
- Team honored during 9 of 15 quarters as "Top Producers"

Assigned the added responsibility of Verifier (09/96 to Present):
    Responsible for the verification of all sales. Extensive customer service and
    problem resolution responsibilities.
Accomplishments:
- Only Team Leader within company to double as a "Verifier"

REFERENCES AND FURTHER DATA UPON REQUEST

98—TOUR DIRECTOR

CHRISTINE MICHELLE DYSON
178 Northstar Street
Fairbanks, Alaska 99775

Highly Experienced, Award-Winning Tour Director
Seeking...
EXECUTIVE-LEVEL POSITION WITH TOP-PERFORMING ALASKAN TOUR COMPANY

Professional Overview

A successful track record ... of more than nine years as Executive Tour Director in fast track, high-visibility positions with top-rated international companies. Developed regional reputation for coordinating and directing top-of-the-line, quality-driven excursions throughout Alaska (up to 180 persons per tour). Recognized for outstanding organizational skills, creative programming, public speaking and presentation expertise, and the ability to consistently exceed customer expectations in a highly profitable manner. Work closely with upper management in program development, modification, and enhancement.

Scope of Experience

PRINCESS TOURS CORP., Anchorage Office, Alaska   1992 - Present
Executive Tour Director

Work closely with senior management in developing a wide range of up-scale Alaskan tours for the largest tour company in Alaska. Personally direct nine months of tours, hosting 3,000+ guests annually, totally responsible for the following:

Travel connections	Managing finances for all tours
Hotel and meal arrangement/confirmation	Individual tour coordination
Scheduling connections and transfers	Cruise/tour director while on ship
Handling problems and complaints	Documentation and record-keeping
Introducing new and innovative programs	Anticipating and resolving problems

Ensuring 100% guest satisfaction

Developed extensive contacts and extraordinary relationships with individual tour directors throughout Alaska to enrich guest's vacation experience including Fairbanks, Anchorage, Juneau, Skagway, Ketchikan, and Denali National Park. Led outback tours throughout the Interior, the Arctic, and the Bush, as well as the Yukon in Canada.

Awards or Recognitions:

Glacier Award, 1997 and 1998	Iced Bear Award for Tour Excellence, 1992 - 1994
Featured guest on TV 6 (Anchorage - four times)	Featured guest on TV 12 (Fairbanks - three times)

HOLLAND- AMERICA / ALASKAN TOUR CORP., Anchorage, Alaska   1989 - 1992
Tour Guide

Conducted daily tours for number one Anchorage tour company servicing over 600,000 visitors in a seven-month period. Led half day and full day tours in and around Anchorage working with diverse groups and cultures.

Education and Qualification

Associate in Travel Science, 1988   Fairbanks Community College, Fairbanks, Alaska

* Certified by the State of Alaska	* Dale Carnegie Public Speaking, 1995
* Registered Agent for the Cities of Anchorage and Fairbanks	* Dale Carnegie Public Speaking Advanced, 1998

- References and Supporting Documentation Furnished Immediately Upon Request -

> *Author's favorite.*
>
> *Very attractive and elegant layout.*
>
> *Unique first-person narrative in Professional Experience section that actually works in this example.*

Dianah Ryder

2251 Red Horse Trail   **Home: 913-555-5794**   **Mobile: 913-555-8371**   Marshall, Texas 78240

Education

Associate in Applied Science - Veterinary Technology
Cedar Valley College, Lancaster, Texas - 1993

Credentials

Registered Veterinary Technician
Texas State Board of Veterinary Medicine - 1993

Professional Qualifications

Accomplished paraprofessional with career focus on large animal veterinary treatment facilities.

- Highly motivated, professional and articulate. Special knack for client assistance and education.
- Enthusiastic, positive and patient-driven.
- Skilled and practiced in the following:
  - **Triage.**
  - **Laboratory procedures.**
  - **Surgical preparation and assistance.**
  - **Patient monitoring.**
  - **All areas of clinical treatment procedures—on site and on farm call.**
  - **Dental prophylaxis.**
  - **X-ray procedures.**
- Innate ability to communicate trust and reassurance to sick or injured animals.
  - Stimulate cooperation during treatment.
  - Accelerate healing process.

Professional History

Marshall Animal Hospital	Marshall, Texas
Veterinary Assistant/Technician	1995 to Present
Marshall County Veterinary Clinic	Marshall, Texas
Veterinary Assistant/Technician	1993 to 1995
Hunt County Veterinary Clinic	Greenville, Texas
Kennel Worker/Vet Assistant	3/91 to 9/92

Professional Associations

North American Veterinary Technician Association
International Arabian Horse Association

Personal

Avid fan, student and promoter of legendary Horse Whisperers' training methodology. Studied under renowned trainer Ray Hunt. Promote "Whisperer" training methods by scheduling and conducting clinics throughout the North Texas area.

—Related during client interview

Professional Experience

Critical Care

I remember one particular dog—a black Labrador with the sweetest eyes I'd ever seen—who was suffering from frequent grand mal seizures.

The Doc and I stayed with our patient for 3 days - 24 hours a day - searching for the right combination of meds (anticonvulsants) that would decrease the frequency, duration and severity of the seizures. (The patient was suffering severe attacks every 12 minutes at the most critical point.)

After we got the dog to a manageable level for quality of life, I'm not sure what made me happier—a thank you card from the owner, or knowing that the dog would have a happy life as long as he stayed on the meds that finally worked for him.

Initiation

I was new on the job, and still pretty "green", when the Doc asked me to go out back and load up a bull for one of our clients. This animal had prevailed against other handlers and staff who had been trying to load him—without success—for most of the day.

It was just about closing time. I was tired, edgy and all the other stuff that doesn't work with bulls when I went out to the pen—a pit of freezing mud—and set to work opening an alley to push the bull through. I got stuck! Stuck in the mud up to the tops of my knee-high rubber boots! I was in a small holding pen with a 2000 lb. Beefmaster bull—and couldn't move an inch—seemed to dawn on the bull and me at the same time. It got "kinda western" for a minute...then common sense took over and I scrambled through the fence. I looked back and saw the bull charge my empty boots—still standing upright in the bog.

I got the bull loaded, but I did it from outside the pen. I'd been about as entertained by that bull as I cared to be.

Job Satisfaction

Most of a VT's job is not real pretty. If you work for a large animal vet, you persevere through freezing rain and mud, 100°+ heat, and a lot of cleaning up. But seeing an animal recover from a devastating disease or trauma makes it all worthwhile to me. I'll always remember the first time the vet told me to pull on a calf's leg—during a difficult birthing—and out he came! His momma cleaned him up and nuzzled him toward the milk factory...me and the Doc just stood in the pouring rain admiring another one of God's miracles. If it was easy, I guess everyone would do it.

100—WAITER

WILLIAM HOLMES, JR.
(719) 555 -9050

115 North Union Boulevard
Colorado Springs, Colorado 80909

Hospitality—Food & Beverage

Wait Staff - Host - Bartender - Trainer

EDUCATION

Undergraduate studies, General Education, Pikes Peak Community College
Graduate, Colorado Food Handling Safety Course
Graduate, Colorado Bartending School (License is current)
Graduate, Customer Service Training Program, Antonio's Restaurante
Graduate, Mitchell High School, Colorado Springs, Colorado [1993]

PROFESSIONAL EXPERIENCE

October 1995
to
Present

WAIT STAFF - TRAINER
BROADMOOR HOTEL, COLORADO SPRINGS, COLORADO
(A world renowned 5-Star and 5-Diamond Hotel and Resort with 4 gourmet restaurants)

- Assigned to all four gourmet facilities:
    (Penrose Room, Tavern, Spencer's and the Golf Club)
- Greet guests and make them feel immediately welcome
- Assist patrons with food and wine selections, including daily specials
- Prepare and serve a wide variety of menu items
- Operate and reconcile a cash fund with high volume transactions
- Responsible for training all new wait staff in procedures

June 1993
to
October 1995

[Part-time
evenings]

WAIT STAFF - HOST
PEPPER TREE RESTAURANT, COLORADO SPRINGS, COLORADO
(A small, up-scale, fine dining facility with a high-volume of repeat customers)

- Required to greet all repeat guests by their names and honorific
- Menu presentation is 100% verbal, with kitchen creating special requests
- Extensive table-side presentation of salad and main course items
- Serve as Host on weekends, greeting guests and acting as maitre-d

June 1993
to
Present

[part-time days]

BARTENDER
PATTY JEWETT GOLF CLUB, COLORADO SPRINGS, COLORADO
(A fast-paced and high-volume public facility with a very casual atmosphere)

- Prepare a wide range of mixed drinks from well to call labels
- Significant problem resolution responsibilities
- Order, stock, control and inventory all liquor and mixes
- Troubleshoot beer-wine taps plus refrigeration and dish washing equipment

References and Amplified Background Information Available Upon Request

101—WRITER

Jessamyn Swanson   Writer

Professional Profile

Extensive background in written communication and editorial administration.

Write a weekly small-business feature for Newsday, the country's seventh largest newspaper.

Serve as assistant editor and regularly contribute topical articles for the lifestyle magazine Prestige.

Develop and pitch story concepts, land the interview, do the research, and get the job done on-time.

Writer

cover stories, features, essays, interviews, film and music reviews, sports, fiction, poetry

Editor

content, editorial board management, production, final copy edit, deadlines

Interviewer

newsmakers, politicians, entertainers, human interest

Career Development

ASSISTANT EDITOR AND FEATURE WRITER

Prestige Magazine, a Times-Mirror Publication, Melville, NY 1994 to present

Handle multiple responsibilities using excellent writing, editing and time-management skills. Conduct general copy editing and reporter interface on a routine basis. Assign stories, development, set-up contracts, and establish deadlines for fifteen free-lance writers and contributing editors.

Generate story ideas, conduct interviews, research and write full-length features, Q&A format articles, and celebrity, business and social profiles. Have interviewed Chevy Chase, Bernadette Castro, Jim McCann, Christy Brinkley, Bob Wallach, Robert Klein, Liz Smith, Donna Karan, Alan King, Carolyn McCarthy, Shirley Strum Kenny, Carol Baldwin, and Mario Buatta among others.

WEEKLY SMALL BUSINESS FEATURE Writer

Newsday, Melville, NY   1996 to present

Seek out and research potential businesses, pitch ideas to editor, and stay two weeks ahead of deadline. Maintain diversity of profiles and relate content to current business, political or social events.

Have profiled over 100 businesses since the feature's debut. Originally approached by Newsday to write one article; assignment developed into a weekly profile of unusual entrepreneurial businesses.   One of only a few free-lancers to maintain a Newsday business section weekly feature.

Education and Activities

Bachelor of Arts in English, The State University of New York at Stony Brook, 1992

Jessamyn Swanson   10 First St., Patchogue, NY 11772   (516) 555-5555   JAS@aol.com

INDEXES

Index by Industry/Job Title

General Index

About the Authors

Jay Block, CPRW, internationally certified career coach and resume strategist, is the contributing cofounder of the Professional Association of Resume Writers (PARW). He helped develop the PARW national certification process and is a widely respected national speaker, author, and career coach. The Professional Association of Resume Writers (PARW) is the official organization governing resume standards. PARW aims to elevate the skills of resume professionals. It provides the only certification process for resume writers.

Michael Betrus, CPRW, is the coauthor of *The Guide to Executive Recruiters* (McGraw-Hill), the most comprehensive directory of its kind.

Jay Block and Michael Betrus are coauthors of the bestseller *101 Best Resumes* and *101 Best Cover Letters.*